A Cost Based Approach to Project Management
Planning and Controlling Construction Project Costs

A Cost Based Approach to Project Management: Planning and Controlling Construction Project Costs introduces early-career architects, construction managers, civil engineers, and facility managers to the essentials of delivering projects on-time and at cost. Drawing on the author's decades of experience managing marquee building and infrastructure projects around the world, this primer offers busy professionals a crash course in budgeting, cost estimating, scheduling, and cost control. Chapters break down the details of cost elements, structuring project costs, and integrating budget with schedule, providing novice project managers with the key skills to plan and execute construction projects with confidence and precision.

Features:

- Illustrates the principles of project management and the essentials of cost planning and control with easy-to-understand examples from the construction industry
- Includes step-by-step details of project planning, cost estimating, and management processes
- Offers clear, cost-based methods for defining scope, preparing bids, and planning for contingencies, as well as monitoring progress and determining when to take remedial action
- Contains a user-friendly guide to project management acronyms and terminology
- Provides sample construction schedules, budgets, and progress report forms

An ideal resource for self-study, on-the-job training, or courses in construction, architecture, or civil engineering project management, *A Cost Based Approach to Project Management* makes a worthy addition to the aspiring project manager's reference shelf.

Mehmet Nihat Hanioğlu is a civil engineer whose four-decade career portfolio includes heavy civil construction, residential development, mixed-use hospitality design/build projects, and claims and project management consulting. He has overseen major challenging development projects around the world, including the Hyatt Regency Bishkek, the Wynn Hotel and Resort Macau, and the High Point Terminal in Riyadh. As an instructor, he has shared his expertise with emerging project managers at City University of New York–Lehman College, Illinois Institute of Technology–Chicago, and California State University at Sacramento.

A Cost Based Approach to Project Management

Planning and Controlling Construction
Project Costs

Mehmet Nihat Hanioğlu

Routledge
Taylor & Francis Group

NEW YORK AND LONDON

First published 2023
by Routledge
605 Third Avenue, New York, NY 10158

and by Routledge
4 Park Square, Milton Park, Abingdon, Oxon, OX14 4RN

Routledge is an imprint of the Taylor & Francis Group, an informa business

© 2023 Mehmet Nihat Hanioğlu

Library of Congress Cataloging-in-Publication Data
Names: Hanioğlu, Mehmet Nihat, author.
Title: A cost based approach to project management : planning and controlling construction project costs / Mehmet Nihat Hanioğlu.
Description: New York, NY : Routledge, 2022. | Includes bibliographical references and index.
Identifiers: LCCN 2022003654 (print) | LCCN 2022003655 (ebook) | ISBN 9781032001005 (hbk) | ISBN 9780367776459 (pbk) | ISBN 9781003172710 (ebk)
Subjects: LCSH: Building—Cost control. | Construction industry—Costs. | Construction industry—Cost control. | Building superintendence. | Project management.
Classification: LCC TH438.15 .H36 2022 (print) | LCC TH438.15 (ebook) | DDC 692/.8—dc23/eng/20220401
LC record available at https://lccn.loc.gov/2022003654
LC ebook record available at https://lccn.loc.gov/2022003655

ISBN: 978-1-032-00100-5 (hbk)
ISBN: 978-0-367-77645-9 (pbk)
ISBN: 978-1-003-17271-0 (ebk)

DOI: 10.1201/9781003172710

Typeset in Goudy
by Apex CoVantage, LLC

Thank you Kathleen;
I could not have done this without your support.

Contents

Introduction   69

Project Planning Process   69

   Initiator of the Project Planning Process: Scope   71

   Detailing Scope by Breaking It Down to Tasks: Work Breakdown
       Structure (WBS)   72

   WBS Should Cover the Entire Scope   73

   Example: Dinner Party Project   73

Compiling Detailed Task Information: Work Package Dictionary (WPD)   75

Structured Project Cost: Project Cost Breakdown (PCB)   76

PCB and Project Price Breakdown   80

Different Perspectives for PCB   80

WBS Examples for Different Stakeholders   83

In Summary   83

6   Cost Estimating   85

Introduction   85

Cost Estimating in the Construction Industry   85

Types of Estimates and Estimate Accuracies   86

Cost Estimating Process for Line Items   88

   Quantity Take-Offs (QTO)   89

       QTO for Materials   89

       Labor and M/E QTOs and Durations   89

Estimate Preparation Duration and Accuracy   90

Expert Opinion in Estimating Costs   91

Step-by-Step Cost Estimating Process   91

Project Delivery Systems, Contract Pricing, and Cost Estimating   93

Technology and Reference Data for Estimating   94

Updating WPD With Estimated Cost Data   95

WPD and Project Resource Schedule (RBOQ)   95

Example: Hotel Building Superstructure   97

   Quantity Take-Offs, BOQ, WPD Update, and Project Resource BOQ   97

   STEP 1. Prepare (or Check if Available) Quantity Take-Offs   97

   STEP 2. Update the WPD for This Activity   97

   STEP 3. Assign Resources and Determine Durations   97

   STEP 4. Calculate the Cost by Applying Unit Costs to the
       Quantities   100

   STEP 5. Check Cost and Schedule Acceptability   100

       Option 1. Increasing Crew Size   103

       Option 2. Keeping the Initial Crew Size and Paying Overtime as
           Needed   104

9   Controlling Cost

Abbreviations

A/E	Architect or Engineer
BOQ	Bill of Quantities
BOT	Build Operate Transfer
CAC	Cost at Completion
CF	Cash Flow
CM	Construction Manager
CRV	California Return Value
CTC	Cost to Complete
DB	Design Build
DBB	Design Bid Build
DBOT	Design Build Operate Transfer
EPC	Engineer Procure Construct
G/C	General Contractor
Gb	Gigabyte
GMP	Guaranteed Maximum Price
M/E	Machinery and Equipment
MBO	Management by Objectives
O/H	Overhead
PCB	Project Cost Breakdown
PM	Project Manager
PPP	Public Private Partnership
QTO	Quantity Take Off
RBOQ	Resource Bill of Quantities
WBS	Work Breakdown Structure
WPD	Work Package Dictionary

Preface

The modern project management concept has been around for over six decades and still finds extended use by businesses in several industries. In the past two decades, the number of large, medium, and small businesses that use project management has substantially increased. From financial institutions to healthcare facilitators, space programs to small start-ups, project management has become the preferred means to conduct business. Organizations expect to accomplish ambitious goals and still save time and money by simply labeling every possible activity as a "project". Many routine tasks are named "projects" and then populated to "portfolios". One reason behind naming a routine task a project is the expectation that if anything is labeled as a project, the project managers will accomplish objectives, deliver qualitative and social benefits, and still stay within their cost and duration limitations. This practice generates a demand for "project" managers; managers who can manage any kind of projects.

One popular practice to fulfill that demand for project managers is educating interested individuals a generalized process-based approach to project management. This approach teaches prospective project managers *how to manage a generalized—one fits all—bureaucratic process* rather than demonstrating *how to manage to accomplish specific goals of a project*. In an effort to include all sorts of projects from all sorts of industries a project manager may end up managing, the leading current educators have created a system of broadly defined concepts and terms that are intended to be *tailored* later in the process as per the specifics of respective projects. This creates a dilemma of its own. This widely generalized, prescriptive approach which is not industry and/or business specific, falls short on providing solutions to managers' everyday challenges. Who will be doing that tailoring? If the student is capable of identifying project specifics and knows how to deliver them, most probably s/he would not need to be educated on the bureaucracy of the process. On the other hand, if a student needs to be educated on the process, it would be unfair to expect that student to already know how to identify and deliver the very specifics of projects of all sorts.

Organizations are primed for ambitious accomplishments by simply labeling unstructured activities as projects, and then they expect project managers to deliver poorly defined and/or unrealistic goals of those projects. In addition, project management professionals find themselves in positions to deliver those unrealistically ambitious accomplishments by using processes which are too generalized and far-fetched from the basics. When facing project specifics, the prescribed bureaucratic solutions require judgment calls to be made by more experienced project managers which undercuts the expected efficiency. Engaging several project managers with varying levels of experience does not make business sense for many organizations that are eager to use project management as their preferred business practice.

Construction industry differs from other industries since construction projects fulfill all requirements of the definition of a project. For over eight decades construction projects have contributed to advancements in project management. Traditionally construction contractors are required to deliver a scope, within a predetermined time frame and for the lowest possible price. When the scope and duration is given and the lowest bidder gets the job, the contractors have to focus on preparing their lowest prices. The preparation of the bid price is their first step in planning for a project. Since the contract price is fixed, contractors start projects by scheduling, which is falsely presumed to be their first planning action. The number of alternate delivery systems developed, especially in the private sector, has shifted owners' priorities from only cost to scope, duration, quality, and risk and conflict aversion. As a result, project management priorities, especially in industries other than the construction industry, have shifted from scope, cost, and time management to managing a series of terms followed by the general term *management*.

This book aims to bring the focus back to the basics of project management and presents specificity to project cost, one of the three basic project defining tenets. Project cost preparation is taken as the starting point for project planning as an alternative to the common practice of starting project planning by scheduling. A case is made that both cost elements of activities need to be first planned and then controlled to accomplish not only the cost but also the scope and the duration of a project. Cost planning and controlling is presented as an alternative to "cost accounting". Planned material, labor, machinery/equipment and other resource data derived from the details of activity cost elements is used as references for controlling cost, scope, and duration for selected critical activities and ultimately for the project.

Examples and best practices are purposefully selected from the construction industry due to their comprehensiveness, details, specificity, and transferability to other industry projects. While this book is targeted to emerging professionals in construction engineering and management, its emphasis on controlling costs will be of interest to project managers from other engineering and business areas.

Mehmet Nihat Hanioğlu
Sun City West—Arizona
December 2021

1   Defining a Project

Introduction

Project management has been around for over five decades. It has been used in several industries and still finds extended use in many businesses. In the past two decades, the number of large, medium, and small businesses that use project management has substantially increased. Operating by managing projects has become a common practice for businesses of all sizes. From financial institutions to healthcare providers, space programs to small start-ups, project management has become the preferred means to conduct business. Organizations expect to accomplish ambitious goals and still save money and time by simply labeling major activities as "projects". Many activities are considered to be projects where they are not. One reason for that call is the expectation that if they are labeled as projects, they will deliver their objectives while staying within their cost and duration limitations and meeting or exceeding demanded qualitative and social targets. The practice of erroneously generated projects brings along the demand for "project managers" who are expected to deliver either unestablished and/or wrongly established goals.

Wrongly labeling an activity as a project is one of the most common errors in project management. Since project managed endeavors deliver results within specific time and cost limitations, it is no surprise that conducting businesses through projects is widely popular. However, simply calling some activity a project cannot guarantee delivery of expected results. If the subject endeavor is not a project, it cannot be project managed. A project is a unique endeavor to accomplish unique goals. That goal is a result, a product, or a service. A project accomplishes its goal by delivering its scope for a pre-agreed cost and within the duration acceptable to all stakeholders. Specific scope, cost, and time are the basic definers of a project. Other requirements and limitations can be added to these primary project targets only if their progress is objectively measurable. The quantitative nature of triple constraints enables objective measurement of progress and evaluation of results. Unless explicitly specified and agreed on by all, qualitative requirements that may be included as additional goals may not lead to satisfactory results for all stakeholders. This chapter also presents typical stakeholders, key project participants, and their roles. It further discusses project delivery systems and contract pricing options and how coupling them collectively governs project success.

DOI: 10.1201/9781003172710-1

Project Definers

Triple Constraints

The most prominent project definers are project scope, cost, and time. Traditionally these three are collectively named as the *triple constraints*. They are presented as limitations of what will define a project.

The scope constraint defines what will be included and to what extent that work will be performed. The cost constraint sets the limits to resources to be consumed and the time constraint sets the time window during which both the scope and the cost requirements need to be accomplished.

A scope developed to achieve a target "*at any cost and at any time*" is not a project; it is a discretionary endeavor. However, a scope to achieve a target "*at any cost and as soon as possible*" is a rare occasion but it is a project. The first one of these two cases cannot define either a cost or a time. "At any cost" does not qualify for a project definer when coupled with another open ended "at any time" time requirement. Whereas when one of these requirements is definitive, they may not make the other become definitive but in combination they sufficiently define a project.

Scope

Projects are endeavors with specific targets. A project can be broken down into tasks with individually differing targets which collectively enable accomplishing project targets. The targets of a project can be a result, a product, a service, and any combination of these three. Examples of results are the outcomes of research or an experiment; examples of products are developing a building or a microchip, and examples of services are an art design or a plant calibration.

Whether specifically outlined or not, targets usually have a cost and time dimension, a requirement, a limitation attached to them.

What needs to be done, what needs to be included/excluded and how it needs to be done to accomplish those targets constitutes the *project scope*.

Scope defines what needs to be done to accomplish project goals. It is the definition that includes all parts of the project. In some cases, scope also defines what the project excludes. Scope is not a *specification*. Scope defines *what* and specifications define *how*. A specification defines the function, performance, capacity, quality, and features of products, results, and services. Specifications may also include specific techniques/methods to be implemented in accomplishing the work included in the scope.

A simplified scope statement (i.e., scope of works) for a house would be:

> "A single-story house of 2,000 sq. ft., to be built on an owner provided lot in the X suburb of Y city of Z state as per Owner provided architectural and engineering design documents and specifications. Exterior hardscape and landscape is excluded and the Owner is expected to obtain all permits."

The design documents and specifications are prepared to provide all details for all parts of the scope such as structural and architectural elements of the house as well as all equipment and appliances.

Specifications are necessary only if what is specified is included in the scope. Scope defines the project and specifications do not. In other words, scope can define what the project is without any specifications but specifications without a scope cannot define anything for the project.

Without a target, there cannot be a project.

Cost

To accomplish their goals, projects consume resources in several forms such as material, labor, machinery, equipment, technology, cash, loans, know-how, and all other kinds of assets. The project cost, which is the total of the entire resources required to complete a project, is most commonly expressed in monetary terms.

Cost is a limitation included in the project definition. Theoretically cost on its own does not define a project. It is one of the requirements, one of the project goals that supplement the project scope.

Cost may be a maxima/minima effort or may come in another form of optimization. In traditional competitive bidding the owner targets to obtain the minimum cost for a pre-defined scope. Maximum target (i.e., not to exceed) price contracts aim to cap the price for a specific scope. Build-Operate-Transfer (BOT) and Design-Build-Operate-Transfer delivery (DBOT) systems focus on the duration from the start of the project to the transfer of the product to the owner. In this model the owner does not *pay a contract price* to the developer. The price of the project is the *duration* during which the owner does not collet revenues from the product and grants that right exclusively to the developer.

The objectives of *not to exceed* (i.e., guaranteed maximum price) contracts are simply getting the most bang for the buck without breaking the bank. "Getting the most" can be in quantitative and/or qualitative form. Take the project example of developing a smaller dimensional size for a computer chip having a bigger processor capacity for a project cost not to exceed a predetermined amount. While the size reduction and processing capacity increase are examples of physical quantity maximization/minimization goals, this project may also include qualitative objectives such as maximizing use of more environmentally friendly materials and minimizing pollution/carbon impact on the environment.

If the exemplified project were not supplemented with a price cap, it would end up being cost wise an open-ended research and not a project.

DBOT projects present good examples for projects supplemented by a cost not defined in monetary form. For example, a developer would finance, design, and build a commercial property on the owner's lot for the rights to operate and lease the developed property for a contracted duration. At the end of that duration those rights are terminated and the property is transferred to the owner. In this model the developer finances the project cost and recuperates that cost and a profit through leasing/operating it for the period of the contract. The owner provides the scope of this project in its most general form and expects that scope to be enhanced by the developer. In return the developer builds up that scope by using expert designers, builders, and operators for maximized business benefits.

In this example, the price of this project is determined with the contracted project duration. The less the contracted duration, the better the price for the owner. The developer of this example has a different challenge. Since the contract duration is all inclusive, the more time it takes to design and build this property, the less time will be left to collect revenues from operating and collecting revenues and hence the project will be more costly. The developer is challenged to minimize its designing and building and maximize the operation duration.

Although very rare, some projects may be initiated with no cost limitations. Such projects have other priorities such as national and/or personal pride, distinguishment, and recognition. However, even those exceptional projects that start with "*whatever it takes*" cost target should be expected to encounter some sort of accountability somewhere along the line. President Kennedy's *going to the moon* project can be an example for such a case. This project was initiated with no cost and time limitations but naturally there were expectations.

In short, a project cost comes in different resource types and it provides depth, expectations, and limitations to the scope. Definition of a project is incomplete without a cost associated with it.

Duration

Projects have definitive start or end dates or durations associated with these dates. The combination of any two of these variables define the third one. "As soon as possible" is one common start date or duration definition and is a valid one. "No later than . . ." is another common time requirement for a project. The start and end dates and the duration are not sufficient to define a project on their own but without them the definition of a project is incomplete. Some goals, put out as to be performed *whenever and however long it takes* is a wish list item and not a project.

The start and end dates for a project may be chosen arbitrarily but they are mostly mandated by the higher objectives of the project. A higher objective of a project may range from a social benefit to a business and/or personal benefit. If the owner of a building project is a government agency, the benefit of this building is expected to be some public benefit from the operation of it. An individual may own a building project for private use of it. A business may invest in a building project for some business return through the use of or operating that building. In all these cases an early delivery would be preferred but a delivery date may be determined by optimizing the design and construction duration of the building. If that project happens to be a monument for some commemoration at a certain date, then the end date determines how long it can take to complete the project.

Chicago's Millennium Park is a good example of projects with targeted completion dates. This great project was started in 1988 to be completed in 2000 for Millennium celebrations. With several changes to its original scope, related design, and construction, the opening celebrations could only take place in 2004. This project is also a very good example of how projects can be successful although they breach time and cost constraints in order to accomplish a higher priority.

Time and Money Complications

"Remember that Time is Money" wrote Benjamin Franklin in 1748.

Not much has changed on that front since then. Time is still money, especially from a perspective of defining projects. The time and money relationship is a complex relationship. Duration may dictate projects to have higher or lower costs. On the other hand, the amount of dedicated resources may shorten or lengthen project durations. But this is not a direct relationship. Sometimes more time may reduce cost or more time may cause cost to increase. Likewise increased resources may reduce duration or after a critical point adding more resources may not have any effect on duration.

A quote from Warren Buffet reads:

"No matter how great the talent or efforts, some things just take time. You can't produce a baby in one month by getting nine women pregnant".

Crashing critical path activities is a good representation of cost and time interrelationship. *Crashing* is the effort to decrease project duration by decreasing some critical activity durations (i.e., crashed) by adding resources (i.e., adding cost) to them. This process is repeated until several project durations with corresponding costs are obtained until adding resources does not result in meaningful duration gains. Then a choice is made based on priority. If the cost is the priority, then the lowest project cost determines the project duration. If the project duration is the priority then the cost that corresponds to the shortest duration can be selected. The third option is choosing a project cost and time both of which are within their constraints.

Project cost being comprised of different types of costs, some of them time dependent and some not, is another factor that complicates project cost and time relationship. The composition of types of costs that make up the project cost and their variations with time requires a careful analysis to produce meaningful and credible conclusions that can be utilized in finalizing project cost and time.

Do the Triple Constraints Sufficiently Define a Project?

The triple constraints minimally define a project. Without them there is no project. By using these main project definers one can create several so-called projects which actually do not qualify as projects. It is even possible to generate a collection of such so-called projects; *portfolios*.

However, an endeavor requires more than that to be rightfully defined as a project.

An example is *a "daily dinner project"*. This project's target is to feed yourself. The scope is a main dish accompanied by some side dishes and a dessert. The cost limitation is \$35 and the time limitation is 1 hour between 5:30 to 6:30 pm.

What is described in this example may be considered as a project and a *portfolio* can be populated with some variety like "hamburger dinner project", "pizza dinner project", and so on. The good news is most probably you will not be required to be *certified* to manage this portfolio.

Obviously, this example does not represent the intent of a project definition although it has a defined scope with cost and time constraints. This is a routine activity with some options. It would have been a much different case if that "dinner project" was about a special anniversary dinner.

So what is missing here? We need more project definers.

Uniqueness

A project is a unique endeavor defined by a combination of unique requirements and it has unique goals. Although some may come very close, it is extremely unlikely for a project to repeat itself. Consider a building project. If two of the same buildings were built simultaneously, the project is building the two buildings. It is not two exactly the same building projects side by side. If one building is built first and after a while the second building is started, the second building should be expected to be very close to the first one but never exactly the same. Weather depending on the season, other climatic conditions, team members, labor and M/E resources, market conditions, even the soil conditions are some of the things that should be expected to be different. Any one of these factors and many others not listed here may have a dramatic impact on the second project.

Limited controllability of the project work environment is another property that differentiates a project from other endeavors.

One-Time Event

Projects happen only once. They are not expected to repeat themselves, not even for a second time, although they may come close to being the same. However, the tasks that constitute the project may repeat their own processes multiple times. Imagine building a hotel property. The construction of this building including its furniture and furnishings is the project. While some of the involved work could be unique, some of it is repetitious. The marquee sign at the property entrance is one task in this project and so is the stone flooring throughout the building. While the task of designing and building the marquee sign can be categorized as a project within a bigger project, laying down the stone floors is an activity. The main difference between these two tasks is that one of them is a one-time event whereas the other repeats itself in many different areas of the building.

It is not uncommon that some projects may not have a precedent. Unique designs to create recognition and separation from competitors introduce first time ever endeavors, making planning and execution even harder for these projects.

The good news is when broken down, these projects may comprise tasks that are well known to estimators and planners.

Comprised of and Breakable to Tasks

A project is a combination of several interrelated tasks. A task is the most detailed part of a project in a family tree type of a breakdown. Projects are broken down to tasks so that these tasks can be estimated, scheduled, and assigned for execution. Tasks have relatively lower costs and shorter durations but still require some planning, organizing, leading, and controlling in order to get accomplished. While it may not be technically correct to call tasks projects, there is great benefit for emerging managers to treat their assigned tasks as projects by properly applying management functions to them. Some tasks may be treated as *sub-projects*; a smaller project within a project, and some tasks are not any different than daily routine duties.

Consider a team member—a procurement officer—tasked with shopping for and procuring the daily consumables for a project. Although many characteristics of this job may seem to be similar to those of a project, this officer is not managing *daily projects*. This officer is just accomplishing tasks. Actually, it is irrelevant that this officer is a project team member; this position can take place in any other type of operation and/or organization.

Task Interdependence

The tasks that make up the project are interdependent. This interdependence may cause failure of one or more tasks when one precedent task fails. One failed task may cause the entire project to fail. Consider a wrong batch of concrete poured for one floor of a multi-story building. This may be the end of that building project or the project may suffer major deviation from its original completion targets. If a similar mistake took place in a production line, say a manufacturing operation produces some metal parts using the wrong metal mix, the recovery would be relatively easier and the overall production may not be fatally impacted.

Other Requirements and/or Expectations

There may be numerous other requirements for and/or expectations from a project. Such requirements and/or expectations are not sufficient to define a project but they complement the triple constraints and their implications. If a project will take place in a specific country, at a certain time of the year, the regulations of that country are expected to govern and specific precautions for dealing with the weather conditions specific to that region are expected to be taken even if the project scope may/may not specifically require them.

Safety, specified production quality, local weather patterns, and requirements from authorities having jurisdiction over the project locality are some examples of such requirements.

Expectations are mostly of a qualitative nature and are subjective to individuals involved. Job satisfaction, career development, and work life balance aspects are increasingly included in evaluations of project success.

A different category of expected but not mandated requirement is *ethics* and *social responsibility*. Ethical and social considerations are one generally accepted and increasingly required step in managerial decision making. Although they do not define a project, their inclusion substantially contributes to stakeholders' positions in establishing project targets.

Can Value Be a Project Definer?

As more and more businesses are managed as projects, it is inevitable that these businesses may perceive and implement project management as a *value driven* business process. Not only is value a complex concept, but placing value as a project defining factor and/or as a project deliverable is quite challenging, if not totally impossible.

Value can be a quantitative or a qualitative variable. In addition, quantitative values may as well be subjective just like the qualitative values are. How much a specific object is worth is an example of a quantitative and subjective value. That object may be worth a few dollars to someone and thousands to someone else. As a qualitative concept, value is the product of some shared opinion. Opinions go back to the fundamentals of social structure where "right" and "wrong" leads to defining "good" and "bad" or vice versa.

Unless it is specifically defined and agreed upon by stakeholders as a project target, evaluating a project's success on such a complex concept as value is a good example of setting up unattainable goals and should be avoided.

What Happens If It Is Not a "Project"?

The simple answer is "nothing".

"Nothing" here is neither a negative nor a positive response. It indicates that one cannot accomplish goals by naming endeavors as projects. If accomplishing goals requires management processes, then application of appropriate methods and techniques of those processes will make that possible; and not creating so-called projects, programs, and portfolios and the bureaucracy they are accompanied with.

If project management methodologies were applied to an effort wrongly defined as a project, the results obtained should be the same as what they should be if this effort were simply "managed". The only difference could be the amount of resources unnecessarily consumed to obtain the same results. "Project" management requires implementation of specific cost and time planning and controlling techniques to work in unison, along with organizing and leading functions. If there is no specific cost and time requirement and/or if there is no need for a specific cost and time management, then there is no need to name this effort as a project and try to project manage it; simply managing it should be sufficient to accomplish its targets.

The "*daily dinner project*" mentioned previously is a good example that can be used here again. This effort requires planning but it does not require scheduling beyond making a reservation at most. It requires some cost planning which is not a part of this effort but is a result of our personal budgeting. It does not require much of any other organizing and leadership. So the results quantitatively and qualitatively should not differ much if a very detailed budget and a schedule were to be prepared and coordinated for this "effort".

On the other hand, if this were a special occasion (i.e., wedding, a milestone anniversary, etc.) dinner for close friends and family and you were put in charge to put it together, you might be in a totally different position. The target of this "project" would be to create a memorable dinner gathering for that special date. The scope of this project could include logistics (i.e., rentals of equipment, hiring, and outsourcing), entertainment infrastructure and selecting food and beverage, details of which is provided in Figure 1–1. Clearly this "Dinner Project" will be a special event and will require much more than the routine dinner effort. Being a project, it will require planning of that specific scope to produce the "memorable event" target to take place on that specific date and for the best that can be procured for the targeted budget.

Construction Projects

Humankind has engaged in constructing things even earlier than when the concept of time was acknowledged. Several individuals, tribes, city communities, kingdoms, and empires all built things to serve themselves in one way or another. Some of those constructions served people and made life easier for them by controlling mother nature and protecting them from the elements. Some projects enabled humankind to harness nature's resources, which in return was used to advance prosperity. There were also monumental projects that were built to impress others for thousands of years.

Types of Construction Projects

The construction industry is one of the largest economic sectors in the United States and in the world. In the 2020s, the value of projects constructed add up to over 1.5 trillion dollars in the United States and tops 10 trillion dollars around the world on a yearly basis. These numbers are comprised of projects that can be categorized in several ways.

Public and Private Construction Projects

The most general categorization is by ownership. This categorization separates private and public construction projects. While the actual scope of projects under

SPECIAL OCCASION DINNER

1. LOGISTICS

Select and rent tables, chairs, etc.
Tabletop selection
  China
  Silverware
  Glassware
  Table cloths
  Napkins
  Name cards, candles, flowers
Kitchen & bar setup
Garbage cans
Invitation
  Make list
  Email & confirm
Select & engage
  Entertainer
  Caterer
  Servers, barperson
  Photographer

2. ENTERTAINMENT

Set up stage & dance floor
Sound system & lights
Set up projection screen

3. FOOD & BEVERAGE

Food
  From the Caterer
    Appetizers
    Salad & dressings
    Main course
      Steak
      Fish
      Vegetarian
  From the host
    Cheeses & crackers
    Dessert
Beverages
  Wines
    Reds
    Whites
    Bubbly
  Liquor & Cocktails
  Soda, water, coffee

Figure 1–1 Special occasion dinner project outlined in three subsections.

both ownerships is the same, project processes and delivery are mostly and substantially different.

Buildings: Residential and Nonresidential Building Projects

Both residential and nonresidential projects are building projects. Buildings are constructed of wood, masonry, steel, and reinforced concrete structures depending mainly on how tall they can be feasibly and seismically safely built.

Examples of residential buildings cover single family homes, compounds, mid- to high- rise condominiums and co-ops, and various types of public housing buildings.

Nonresidential buildings can be categorized by their occupancy types. Commercial, industrial, office, educational, healthcare, hospitality, religious, amusement, and recreational are examples of occupancies for nonresidential buildings. Naturally such buildings as well can be low-mid-high rise buildings that are constructed using several types of materials and structures.

Heavy and Civil Engineering Construction Projects

These types of projects mostly involve earthmoving and in situ concrete work. Mostly infrastructure related services and associated structures are included. Roads, railroads, bridges, irrigation and drainage systems, dams, pipelines, aerial and underground transmission lines are examples of heavy and civil engineering projects.

Other specialty projects, such as ports and harbors, waterway dredging, soil improvement—special foundations, caissons, underwater power transmission and pipelines, and tunnels also belong to this categorization of construction projects.

Stakeholders and Key Participants

Stakeholders and key participants of projects are distinguished by how the results of their decisions/performances impact their own interests in the project. *Stakeholders* of a project are subject to financial and/or professional consequences due to their decisions that govern their project performances. *Key Participants* on the other hand, in their advisory and/or regulatory roles, are in no position to suffer either financially or professionally if a project fails due to their performance during the project. Examples of key participants include but are not limited to suppliers, vendors, manufacturers, testing-inspection institutions, all consultants (i.e., business, legal, tax, financial, technical consultants), and all authorities having jurisdiction on the project.

Owners and contractors (i.e., builders under contract) are the main stakeholders of construction projects. The designers and especially the architects are also referred to as construction project stakeholders since traditionally they represent the owners, especially in residential building projects. Naturally the owner carries the overall risk of project performance of all other stakeholders and key participants. The contractor contractually guarantees the performance required by the owner by providing a *performance guarantee*. In addition, most construction contracts include penalty clauses allowing the owner to financially penalize the contractor for delays. Designers and architects are not subjected to such performance criteria and delay penalties. Their only exposure is their professional reputation being associated with a failed project.

Stakeholders' Differing Project Perspectives

The stakeholders may have vastly differing goals and priorities although they are all engaged in the same project.

The owner has the full responsibility for the risks and benefits to be expected from the entire project venture. The primary focus of the owner is to complete the project within the budgeted cost so that the targeted returns can be harvested in due time. The designers' primary focus is on designing the best possible and unique product whereas avoiding cost over runs and delays are the primary goals of the contractors. The operator/manager, if included as a stakeholder, focuses mostly on a high-quality/reliable product that is built to serve the brand's established service flow and business model.

Since these priorities and goals may businesswise clash with each other, it is essential that the owner, being the ultimate beneficiary of the project, take the necessary steps to team up these stakeholders to maximize their potential synergy.

Basic Functions of the Stakeholders

The stakeholders have to perform their own functions while overseeing the functions of other stakeholders and coordinating their own functions with them. Understanding other stakeholders' functions and their priorities is key for successful teamwork and for the success of the project. Some basic functions of main stakeholders are listed next.

For the owner of a mixed use property development, the project includes securing:

- land
- finances
- permits
- facility operator
- designer
- builder

and coordinate the other stakeholders and key participants.

For the designer(s), the project includes:

- Architectural, structural, mechanical, electrical, plumbing, interior design, and other specialty design drawings

- Specifications
- Design coordination
- Construction coordination
- As-built drawings

The project for the builder includes:

- Bid preparation and securing the contract
- Cost and time planning
- Organization and mobilization
- Performing the work including subcontracting, procurement, and outsourcing where /when required
- Progress monitoring and reporting
- Safety, quality, and security measures
- Coordinating bid documents and construction drawings

The Operator as the Fourth Stakeholder

In some projects a fourth party, the operator, may be included as a stakeholder. Especially for properties that will be managed and operated by chain brands, the inclusion of the manager/operator has become essential for owners' business models for such projects. Brand recognition and consistency in offered services have been two main factors securing better returns for businesses. A good example for such projects is hotel development projects. Owners entering management or franchise agreements with hotel chain brands engage them as non-investing or investing partners. Hotel operators on the other hand use design elements to build their brands and offer consistent and well strategized services and amenities for their several tiered sub-brands. When needed, such operator/managers have the ability to tap into their resource pools to support or supplement the properties' existing operating teams with minimal transition/orientation time. Centralized services such as promotions and reservations are more efficiently provided that further advances business returns. As much as inclusion of the operator/manager as a stakeholder improves owner's business, it creates a business dilemma for the contractor. The operator/managers are not a party to the construction and fit-out contracts between the owner and the contractor. The quality and reliability of contractors' work is the priority for the operator/manager since better quality work helps maintain their brands and minimizes down time—maximizes revenues from operations, whereas the contractor enters a contract to fulfil the minimum requirements. In other words, the contractor's price is defined by the contract documents and cannot be kept the same if another stakeholder requires cost increasing changes. This situation can be avoided by engaging all stakeholders to work together before any contractual obligations are fixed. This will allow the operator/manager to understand and comment on the end product quality/reliability concerns they may have and the contractors to establish a good understanding of the end product before they finalize their cost estimation and contract price.

Examples of other construction projects where the operator/manager is a fourth stakeholder are hospitals and airports. Since these properties involve multiple services such as hospitality, food and beverage, and retail along with the specific requirements of the operator for their primary functions, it is essential to have this fourth stakeholder's input in designing, building, and fitting-out such properties.

Project Delivery Systems

The project delivery systems define the roles and contractual relationships among the stakeholders. The delivery system is the choice of the owner either due to legislation or by preference. Not all delivery systems deliver the best for all types of owners and their various types of projects. A private road construction project may have major differences when compared to a public road project because of the differences in the ownership circumstances. The private owner may have the option of inviting and negotiating a contract with a contractor of its choice whereas the public owner may be obliged to have a competitive bid among all qualifying contractors. In addition, public owners may not have the option to choose a delivery system for their projects that private owners do.

Most Common Project Delivery Systems

There are several project delivery systems that are designed to match the organization size, capabilities, and priorities of owners. The following are the most common ones.

- Design-Bid-Build (DBB): Owner separately contracts with designers, builders, and operators and manages the project.
- Design and Build (DB): The owner contracts either a designer or the builder to oversee the design and construction as the general contractor (G/C); may hire a third party to manage the G/C. Operator and major suppliers/manufacturers—if any—are contracted separately. EPC (Engineer-Procure-Construct) delivery is very similar but is used more for engineering heavy industrial projects.
- Construction Manager (CM): Owner contracts a management company to contract and manage the project on its behalf. CM at risk version stipulates a fixed fee or a guaranteed maximum price (GMP).
- Public-Private Partnership (PPP): Owner is a public agency and contracts a private entity to finance and design-build the project. The project is paid for by its operational income for a predetermined period which is the most important bid selection criteria. Build-Operate-Transfer (BOT), Design-Build-Operate-Transfer (DBOT) are two specific examples of PPP delivery methods.

While these represent the basic delivery systems, they can be utilized in combinations, generating an even bigger number of options for owners to choose from.

Construction Contract Pricing Options

There are several contract pricing options than can be coupled with the chosen construction delivery method. It should be noted that although they are referred to as *contract types*, they do not relate to contract content, language, and defined procedures and/or processes that would define types of construction contracts. These options relate to pricing only. These pricing options can be used with types of construction contracts that are offered by international organizations like the World Bank, or professional organizations like American Institute of Architects and by large public or private owners.

Unit Price Contracts

A Bill of Quantities (BOQ), prepared either by the owner or by the general contractor (G/C), provides quantities for specified tasks. The contractor quotes unit prices estimated for each specified and quantified work. The quantities of initial BOQ is expected to have insignificant variations since exact quantities are estimated at the time of the contract and the final quantities will only be known at project completion. Unit Price contracts best serve projects where there are:

- Not many different types of work
- Scope is mostly repetitive work
- Difficulty in finalizing exact quantities

Road construction or pipeline projects are examples of candidates for unit price contracts. The scope of these projects comprise earthworks (i.e., excavation and backfill) and placing layers of asphalt/concrete or laying pipe for lengths which might have slight changes as adjustments are made along the route.

Lump Sum Contracts

One all-inclusive price for a well-defined scope is the agreed contract price for the project. Several versions such as Guaranteed Maximum—or Not to Exceed—Price are options when parties are comfortable with the completeness of scope and these options provide room for slight modifications if any. A Target Price is an option where design/scope changes are expected. The purpose of using a lump-sum contract is to mitigate the risk of significant price changes above and beyond a total which is not feasible for the project.

Lump-sum contracts best serve projects where there are:

- Several types of designed—unique—work
- Not so large work quantities
- Imbalance in quantities and other resources required to carry out work (i.e., a small quantity of excavation requiring the mobilization of heavy equipment)

Building renovation projects are best served with lump-sum contracts where the owners do not need to be involved in details of several hundreds if not thousands of scope details and at the same time, they need to avoid significant price increases.

Variations of Cost Plus Fee Contracts

Cost Plus Fee is the contract pricing choice where the scope definition is too vague for the contractor to commit to a price. The owner guarantees the contractor's fixed and variable costs along with a fee which can be a percentage of the cost or a fixed total. Such incentivized contracts best serve projects where the project is highly design oriented and the design—and accordingly scope—is expected to be developed as the project progresses. A monumental feature for a public area or for a prominent building entrance is an example project candidate for Cost Plus Fee contract pricing.

In Summary

- Scope, cost, and time limitations, the triple constraints, minimally define a project.
- Being a unique and one-time event supplements the triple constraint definition of a project.
- Projects are breakable down to tasks that can be individually managed by several team members.
- There may be several other limitations and/or requirements that a project is targeted to accomplish, value being one of them.
- If set as a project goal, value, being a qualitative variable, may be viewed substantially differently by stakeholders and key participants.
- The construction industry covers both public and private sectors carrying out several types of projects.
- Main stakeholders of a construction project are the owner, the contractor, and with relatively smaller and limited risk exposure, the designer.
- Key participants of a construction project only have advisory or regulatory roles.
- The goals, priorities, and functions of stakeholders and key participants are not necessarily the same.
- Construction delivery systems define the contractual relationships among the stakeholders.
- Several contract pricing options can be used with delivery systems so that project risk can be equitably shared by stakeholders.
- Selecting the most suitable delivery system and contract pricing option for a specific type of construction project is key to that project's success.

2 Management, Project Management, and Construction Projects

Introduction

This chapter offers a general management brief as a refresher for readers who already possess management background and as a further study reference for those who do not. Due to the profundity of the subject, only fundamentals are covered at minimum depth. Managerial concepts, duties, environment, accountability, and dependency are summarized. Skills and styles for different roles at different levels in organizations are mentioned. Decision-making process and common mistakes are outlined. Planning, organizing, leading, and controlling functions are briefed. Especially for those who do not have management backgrounds, the need to learn general management and the common mistakes managers make are emphasized. Historical roots of management, scientific management, general administrative theory, behavioral, quantitative, qualitative, systems, and situational approaches are visited. Project management learning, its current challenges and trends are mentioned as a glimpse into the future of project management and project managers. Typical phases and key activities in those phases of a construction project are presented. The differences between cost and time considerations of an owner's and a contractor's construction project managers are outlined.

Modern Management

Management is getting things done effectively, efficiently, and on time by using several resources while keeping those who are involved happy.

The **"getting things done"** part of the definition relates to the main management process that is collectively planning, organizing, leading, and controlling functions.

Being **"effective"** relates to doing the right thing for the right task. The task may have quantitative aspects such as addressing the right problem and/or may have qualitative aspects such as doing what is ethically right. **"Efficiency"** relates to the ratio of what needs to be put in to obtain the required (or unrequired) result. The input could be any resource such as money, material, time, equipment, software, and similar others. How much we input such elements for obtaining that result and how much we are satisfied with the result is one way of defining

DOI: 10.1201/9781003172710-2

efficiency. **"Those who are involved"** are the teams who are actually conducting the work, the *stakeholders, and the key players*. Depending on the project, the list of "those who are involved" can be extensive. Generally, the project stakeholders are the owner, designers, contractors, and in some projects property and/or facility operators. The key players list includes local and/or federal government agencies, financing/lending institutions, consultants, domestic and/or international awareness groups, and similar others depending on the size and type of the project.

Why Study Management?

Organizations have a never-ending appetite for improvement, for more effective and efficient ways of conducting their businesses to achieve their goals. Several approaches, techniques, methods supported by technology and human studies are innovated and used to quench that desire. Throughout the years of dramatic changes in how businesses are conducted, the need for good people and good management has not changed.

Management practices have been affected by humanities and social sciences and therefore, studying management has been traditionally related to anthropology, economics, philosophy, political science, psychology, and sociology. Currently, management knowledge and skills are minimally required for anyone aspiring to move up to higher organizational positions of any industry.

It can be argued that a well-seasoned manager can *manage* anything since that manager will not be involved in doing the actual work. From that one could jump into the conclusion that a properly trained manager can manage anything even if he/she does not have a specialty background in that specific area.

While that may be a valid argument, it is not easily applicable. The simplified case of a surgery would be a good example. One could even go to the extent that a surgery could be defined as a project—only if the cost could be estimated up front. Do you believe a professional manager could manage that operation instead of a lead medical doctor? The answer is yes but highly unlikely. While it is theoretically possible that a well-trained manager could lead in that operating theatre, it would be quite hard to convince the patient, the medical institution, legal, and insurance entities involved that this is their best option.

But how about that lead doctor and the team being trained in management as well? This option combines the knowledge and expertise of subject specifics with management efficiencies and effectiveness, which is obviously more advantageous than being limited to only one or the other.

One other major discipline that could use management training is engineering. Although a good portion of engineers end up managing projects and/or programs, engineering curriculums rarely include management courses. Instead they prefer including courses that focus on quantitative approaches to optimizing cost and time. Lacking the behavioral and qualitative factors that are proven to have equally important impact on the outcome, engineers find themselves sidelined in processes that involve interactions with other disciplines.

General management study complements the specialist in any field. It brings forward the synergetic potential of teamwork. It increases work life quality by socialization and by preventing introverted tendencies of individualistic practices. Last but not least, management training extends individual employee burn out duration, a sure benefit for both the employee and the employer.

This chapter only summarizes basic management principles and management learning. It is merely a glimpse of several subsections of the subject, each of which take a decent size book of their own to properly cover all their specifics. The reader is highly recommended further reading and training in their field of interest.

Who Are Managers?

Anybody who supports and supervises individuals, teams and/or organizations to accomplish performance goals is a manager. Although managers do not actually do the work themselves, they are responsible for the work they *activate others* to conduct.

Everybody Manages Something

Whether intentionally or not, everybody manages several things in their everyday lives. Be it parents, siblings, kids, or personal relationships, we have to manage having a good relationship with others in our immediate circles. We have to understand our needs and theirs, sometimes negotiate and sometimes give or take what we need so that we can get by. Everyday commute to work or school is another example. We need to pay attention to the weather forecast and traffic news to make sure we are on time to that meeting or class. The point here is that even for the very minor daily activities, we do have a preparation, execution, and as needed correction routine. As stipulated in these examples, everybody acts a manager but they may not be referred to as *managers*.

What Do Managers Do?

Managers get things done. They use all sorts of resources to get things accomplished. To obtain a certain result, they set up goals, they determine what method to use, they organize the resources needed, and they function as the executive of the process. What has been outlined previously can be re-worded as they plan, organize, and lead others to accomplish goals. One other thing they do is they observe these three functions as to what is working and what can be improved. Based on their observations, they make adjustments and corrections. Observing and taking necessary action is managers' controlling function. In other words, managers plan, organize, lead, and control to accomplish certain targeted results. They also implement several practices to support and/or to enhance these four basic functions. Examples of such practices are provided in the Management Practices section later.

Where Do Managers Work?

Managers work in organizations. Organizations are groups of people with a specific purpose who have developed a structure that defines the roles and relationships of its members. Organizations can be as big as the United Nations, an organization dealing with almost every nation, every country in the entire world or, they can be as small as a fundraiser for a local neighborhood team. Regardless of any measurable size related to its operations, an organization is an entity with a specific purpose, which has a systematic structure among its members.

The most common type of organization is "for profit" organizations. Every business falls into this category. Another kind of organization is "not for profit" organizations. Government agencies and charitable organizations are some examples of not for profit organizations. Being a not for profit does not imply that the organization does not have to make money. Regardless of the business type, every organization needs to generate some cash flow to stay active and accomplish its goals.

The work of a manager may slightly differ from one type of organization to another. However, the main duties still remain the same. Managers need to perform the basic management functions, create the organization's support, and maintain the organization's legitimacy and viability.

Management Levels and Titles

Several people work in organizations but they may/may not necessarily be managers. In its simplest form, the members of an organization can be identified by two main categories: operatives and managers. Managers direct and oversee the activities of operatives. Operatives work on the tasks assigned to them by managers and they do not have direct responsibility for any other operative member of the organization.

If the size of an organization allows, managers are hierarchically sub-grouped as top managers, middle managers, and first line managers. In a very small organization there could be only one management position so all levels of managers can be consolidated on that person. In medium to big organizations the titles given to managers define the levels of each manager. Usually Team Leader and Supervisor are the titles given to first line managers. Branch Manager and Division Manager are two common middle level manager titles. Chief Executive Officer, President, and Vice President are among some of the titles given to top level managers.

The managerial hierarchy of an organization can be generalized as the Board of Directors, top managers, middle managers, first line managers, team leaders, non-managerial staff and team members as shown in Figure 2–1.

Accountability and Dependency

Accountability is the ownership of results. Managers are accountable for their organization's performance and for the accomplished results. Accountability is the duty of the manager to report the performance results and assume responsibility

Figure 2–1 Management levels in a large organization.

toward a higher authority. In the simplified organizational hierarchy, first line managers are accountable to middle managers, the middle managers are accountable to top managers, and the top managers are accountable to the board of directors (or trustees). Ultimately, the board or trustees are accountable to the shareholders.

In doing what it takes to get the job done, the managers depend on their teams and organizations. Bigger organizations may have several management levels. Managers of bigger organizations depend on not only the people directly reporting to them but on all of the people who are under their management span as shown in Figure 2–2.

Accountability and dependency of managers create the need for them to support the people they manage. A classical organizational chart for a decent size organization, shows the management levels in the form of a triangle (i.e., Figure 2–2). Since each level is dependent on and accountable for the performance of the levels below them, they need to support those levels so that they can accomplish their set goals. From that perspective the classical triangular organizational

Figure 2–2 Dependency hierarchy of managers in a large organization.

chart should be viewed upside down: boards/trustees have to support the top management; the top management has to support the middle management which in turn has to support the first line managers and the first line managers have to support the non-managerial staff. Serving customers takes place at the very top of this upside down organizational pyramid, indicating that the entire organization needs to support customer satisfaction since they are all accountable for that ultimate goal. Figure 2–3 shows the support perspective being in the opposite direction of dependency and responsibility direction of an organization chart.

Roots of Modern Management

Although people must have managed others throughout human history, the roots of modern management thinking goes back only a few centuries. Eighteenth century advancements in agriculture generated the need for manufacturing better tools for large farms created by wealthy lords buying out smaller farms. The evicted farmers moved to cities to look for employment. The need for tools and availability of workers ushered in industrial manufacturing at larger scales in the 19th century. Consequently, businesses needed higher manufacturing efficiencies to increase production and profits. As a result, improvements in farming and manufacturing improved nutrition, healthcare, and education for societies. However, there were

Figure 2–3 Support structure and direction between managerial levels.

some resulting counter effects such as deteriorated working conditions, decreased wages due to increased labor availability, and an increase in health issues due to deteriorated working conditions and high environmental pollution.

The earliest modern management approaches were introduced to improve production efficiency but neglected the workers as early as the 20th century. These approaches are collectively called the Classical Approach, and included universal, scientific, and bureaucratic approaches as well as five activities defining the management process. Henri Fayol's five managerial duties and practices of 1916 were reconfirmed as managerial functions in 1955 by Harold Koontz and Cyril J. O'Donnell and so far, they are the most widely accepted managerial activities. A counter approach, the behavioral approach, followed with introducing better working conditions for workers as another factor in increasing efficiency. Various other approaches targeted improving the overall process for businesses, employees, societies, and the environment. Table 2–1 summarizes the milestones of modern management approaches. Currently, management scholars and practitioners are challenged with finding situational models that combine parts of several approaches with leadership styles and personalities of people involved. Considering the diversity of organizations, their sizes and objectives, a situational approach is logical rather than adapting universal approach principles. One downside is that the situational approach does not provide specific methods and/or solutions.

Table 2–1 Summarized approaches in management learning

Management Approach	Approach Brief	Pros	Cons
Universal Process	Assumes the same rational process applies to all organizations. 14 principles outline managerial duties and practices.	Management process can be separated into interdependent responsibilities or functions.	Presents *what* should be done but does not offer techniques and methods as to *how* and *why as* other approaches do.
Scientific	Businesses conducted by standards established from systematic observation, experiment, and reasoning; Time—Task—Tool studies; Differential pay for higher production rate; guaranteed day rate (minimum wage).	Promotes production efficiency and reduces waste. Improved working conditions and wages improve production efficiency.	Process disregards the human side of workers. Makes and treats workers as mindless money-making robots.
Bureaucratic	Correcting performance problems by rationalizing organizations by logic, order, and authority.	Well trained workers for clearly defined jobs; careers based on merit; clear organization hierarchy and authority.	Less flexibility and time-consuming process most likely to limit performance, especially in adapting to changes.
Behavioral	Human side of organizations are as important as physical factors and structures. People's needs influence attitude and behavior at work.	Places people, their personalities and needs in the center of organization's success.	Good relations with individuals do not automatically guarantee success.
Operational (Qualitative)	Commitment to continuous improvement and improving efficiency by improving quality.	Doing it right the first time reduces waste of resources and improves brand image.	The goal is a moving target and may become impossible to achieve.

Classic

Table 2-1 (Continued)

Management Approach	Approach Brief	Pros	Cons
Operational (Quantitative)	Developing methods and tools to efficiently transform materials, labor, and technology into designed and desired goods and services.	Improves product efficiency by using statistics, optimization models, and computer simulations.	Lack of focus on quality may cause excessive waste of resources and reduce efficiency.
Contingency	Every unique case requires a situational approach with alternative techniques and methods available to managers.	Situational approach is appealing since it provides alternatives to complex cases.	While asserting there is more than one best way of managing a case, not offering specific solutions does not help practice.
Systems	Unlike other analytical approaches, it favors a synthetic approach to the overall collection of interdependent parts working for a common goal.	Managers see the bigger picture from all aspects; both from inside and outside the organization.	Lacks specific practical tools, techniques, and verifiable facts against appealing concepts and terminology.

Managerial Roles

In performing functions, managers enact several roles. These roles or behaviors have been studied by many scholars. Henry Mintzberg defined the most accepted managerial roles. The ten roles defined by Mintzberg are collected in three main groups:

Interpersonal Roles

Interpersonal roles involve interactions with people within and outside the organization.

- *Figure Head*: performing routine duties (example: signing documents and meeting with customers to address their concerns)
- *Leader*: inspiring and motivating employees
- *Liaison*: engages in connecting teams and team members as well as getting involved in other activities outside of the organization

Informational Roles

Informational roles focus on obtaining, analyzing, and providing information.

- **Monitoring information**: reading reports, maintaining personal relationships and following media to collect information that is relevant to the organization
- *Disseminator*: analyzes collected information and conveys it to the organization
- *Spokesperson*: sharing information that is necessary to inform, update, and motivate the organization

Decisional Roles

Decisional roles include the following four categories.

- *Entrepreneur*: searching for opportunities and implementing any required change so that the organization benefits from that opportunity
- *Disturbance handler*: strategizing for disturbances and crises
- *Resource Allocator*: scheduling, budgeting, and staffing to allocate available resources where they are most needed
- *Negotiator*: represents the organization in negotiations with entities the organization is engaged in doing business with

Managerial Skills

Managers use several skills in their various roles to perform their management functions at their management level and for the results they are expected to attain. Professionalism, critical thinking, teamwork, leadership, communication, and self-management are the most basic skills a manager is expected to have. Commonly, most of these skills are grouped in three categories.

Conceptual Skills

Conceptual skills involve the ability to analyze situations, understand the interactions between the underlying factors, and to make decisions on what actions to take to control that situation in a favorable direction.

Interpersonal (Human) Skills

Interpersonal skills relate to working well with others. It involves understanding, mentoring, and motivating individuals and/or groups. Managers use their emotional intelligence to better communicate, motivate, and delegate to others. Self-awareness and regulation, empathy, and social skills are among the skills that are accepted to be the skills that define emotional intelligence.

Technical Skills

Technical skills enable managers to perform specific tasks that require specific proficiency and expertise. Computer literacy, written and audio-video communications, math, and coding are a few examples of baseline technical skills. Managers usually build technical skills by education, with firsthand experience, and with apprenticeship.

Well known scholar, Clark L. Wilson, also groups the managerial skills in three groups: **Technical, Team Building** and **Drive**. In Wilson's Technical group of managerial skills, **Technical Expertise, Clear Goal and Objective Setting, Problem Solving,** and **Creativity** are included. **Listening for Insights, Directing and Coaching, Team Problem Solving** and **Coordination and Cooperating** are the subsets of the Team Building skill group. The Drive group includes **Standards of Performance** (keeping teams moving toward the targeted new accomplishments), **Control of Details** (overseeing that results meet targets), **Energy** (readiness and willingness) and **Exerting Pressure** (urging the organization to perform as a team).

There are numerous other groupings of managerial skills a manager could use, depending on the situation they are in and at their management level. A manager's ability to establish a network to build a power base and improve his/her positions can be added to the list as **Political Skills.** This skill is of special importance when there is a competition for resources in organizations. **Motivation to Manage** is how motivated managers are to perform and fulfill their managerial functions and roles. Motivation of managers gets stronger as they ascend in organizational management levels.

Last but not least **Self-Management** and **Professionalism** could be added to the list of essential managerial skills. **Self-Management** is the ability to evaluate personal performance against set targets and take corrective action accordingly. **Professionalism** is the ability to instill confidence and credibility in the organization by a strong positive presence and performance.

The use and relative importance of these groups of skills vary according to the manager's level in the organization. Technical skills are more important and they are used more at lower levels of management. The importance of conceptual skills increase as managers advance in their careers and move up to higher management levels. Interpersonal skills (i.e., people skills) maintain their use and importance at all levels due to the fact that the manager's dealings with other people remain the same although the nature of issues may differ at different management levels.

Management Practices

In the continuous circular management process of planning, organizing, leading, and controlling their projects, managers use several techniques and methods to improve results quantitatively and qualitatively. These practices are implemented as sub-functions of the four main management functions. By utilizing these practices, managers:

- Make decisions
- Assure social responsibility
- Act ethically
- Communicate effectively
- Motivate individuals and groups
- Promote innovation
- Make changes
- Solve conflicts
- Develop strategies
- Oversee individual and group behavior
- Build trust

Managers Have to Make Decisions

An organization's success depends on the quality of the decisions made by its managers. Managers make decisions under various circumstances and for several purposes. Plans are doomed to fail if the set targets are unattainable. A **decision** entails choosing one of the alternatives. Reaching to that position to make that choice requires a process, namely a **decision-making process.** The decision-making process involves several steps:

- Identify the problem that requires a solution or an issue that needs a decision
- Identify the decision criteria; the factors that are relevant to the decision to be made
- Generate and evaluate alternatives
- Conduct ethical analysis of the entire process
- Decide on and implement the decision and monitor its effectiveness

An **ethical analysis** of a decision can be conducted by testing it against commonly accepted ethical principles which can be summarized as asking and answering the following questions:

- Long-term self-interest: Does the decision serve the organization's long-term self-interest?
- Personal virtue: Would the decision shame you if it were disclosed to your family, friends, and to the public?
- Utilitarian Benefits: Does the decision benefit the society?
- Individual Rights: Does the decision infringe on any rights of others?
- Distributive Justice: Does the decision harm those who are less fortunate (i.e., the poor, less educated, and unemployed)?

By conducting an ethical analysis of a decision to be made, managers not only choose the most beneficial option but they also choose the right thing to do.

Managers make decisions under several circumstances and by using several approaches. In some cases, a well-rehearsed decision is made for a repetitive issue whereas in some rare cases a unique decision is to be made in a situation never encountered before (i.e., programmed and nonprogrammed decisions). Managers make decisions under certain, risky, and uncertain environments. In theory, two models, namely the *classical* and *behavioral models,* are recognized as decision-making models under such environments. The classical model is applied when the problem (or issue) is well defined and structured and all information, alternatives, and consequences are well known. Managers can optimize their decisions under such conditions. The behavioral model is utilized in an uncertain environment, which is totally the opposite of the classical model; nothing is definitive, there is incomplete information, the problem is not structured, and alternatives and their consequences are not available. Under such conditions the manager is in a position to choose the first alternative that appears to be satisfactory (i.e., satisficing decisions).

Managers use several tools in the decision-making process. Decision trees, break-even analysis, ratio analysis, cost benefit analysis, linear and non-linear programming, queuing theory, and shortest and longest path analysis are some of the methods managers use to generate and evaluate alternative solutions.

Organizations increasingly use groups and/or teams in their decision-making processes. Group members jointly possess different skills, knowledge, and experience, provide a more in- depth look into the problems, and offer more complete solutions by generating and evaluating more alternatives. The group decision-making process takes longer and sometimes suffers from *groupthink*, in which some members withhold their opposing views in order to avoid contention in the group. Nominal Group and Delphi Techniques are widely used group decision-making methods.

Common Decision-Making Mistakes

In an effort to save time in the decision-making process, managers may choose to take judgmental shortcuts called *heuristics.* Heuristics appear in two forms. The **availability heuristic** is jumping to conclusions on readily available information without properly checking the appropriateness of it. The **representative heuristic** is the tendency to assess the probability of the occurrence of an event based on similar other events the decision maker is familiar with.

Escalating Commitment, Framing, and Confirmation errors are other common errors. **Escalating Commitment** refers to staying on a course of action even when the targeted results are not delivered. **Framing Error** is addressing an unverified issue by only considering what is presented. **Confirmation error** is encountered when the decision maker chooses to use only confirming information that supports the decision made.

Management Process

Regardless of their position in the organization, managers are expected to perform four functions. These functions are planning, organizing, leading, and controlling. Several techniques may concurrently and/or separately be used to enhance these basic functions. Figure 2–4 schematizes these four functions of management in respect to several management approaches and managerial roles.

Planning

Planning is setting organizational objectives, establishing a strategy, or selecting a method to achieve those objectives within given parameters and developing a hierarchy of required activities that need to be coordinated. Planning also involves selecting appropriate resources that will be required to accomplish targets. This process includes establishing an overall strategy by estimating activity costs and durations that ultimately define the overall cost and time that would be needed to accomplish goals.

Plans come in several forms serving several management levels. *Strategic Plans* identify long-term goals for organizations. *Operational Plans* (i.e., *Tactical Plans*) set out details to accomplish strategic plans. Strategic plans are developed by and serve the higher management to maintain their oversight for the long-term organizational goals. Operational plans on the other hand mostly serve the middle to lower management to identify their much shorter-term priorities and keep them on track.

Managers use several tools and techniques in preparing plans. Some examples are:

- *Forecasting* uses a probabilistic approach to predicting the future
- *Contingency Plans* identify alternative options for possible adverse situations
- *Scenario Planning* generates options for future scenarios

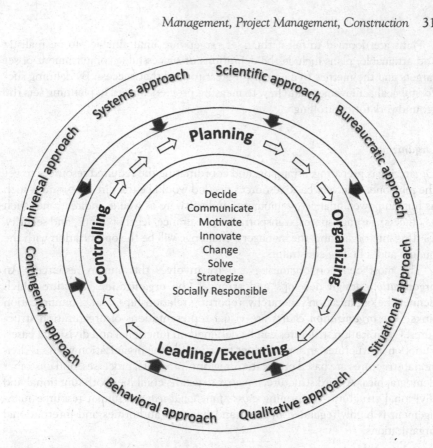

Figure 2–4 Management process and managerial duties.

- **Benchmarking** compares best practices of other organizations to be used in preparing plans
- **Participatory Planning** includes team members to participate in planning to promote ownership and reduce resistance to accepting set goals
- **Management by Objectives (MBO)** integrates planning and controlling by involving managers and subordinates jointly setting goals and reviewing results

The two most common plans are **budgets** and **schedules**. A budget is a plan with assigned resources. Budgets are the summation of costs for all sorts of resources required to achieve the goals *within a time frame*. A schedule is the logical time sequencing of tasks as dictated by the selected construction method. A schedule determines the duration it will take to complete all activities of the job *within a budget*. "Time is money" or "money buys time" are two adages that demonstrate the close relationship between these two distinct types of plans.

Plans are doomed to fail if their set targets are unattainable. To be realistic and attainable, plans include the definitions of successful accomplishment of set targets and the metrics to measure progress toward that success. By defining successful goal achievement and how to measure progress toward it, planning sets the groundwork for controlling.

Organizing

Organizing is preparing, arranging, and coordinating the required resources to get the job done. The very basic resources needed would be the direct resources such as human power, finances, equipment, and software as well as indirect ones such as office(s), storage spaces, transportation, insurance, legal opinions, and security. Getting into organizing, the manager's challenge will be the optimization with the budget and schedule constraints.

The most common organizing activity involves the human resources. An **organization design** develops and arranges the **organization structure** which defines the organization's hierarchy, reporting relationships, and communication links. The **organization charts** are visual representations of organization structures. Organization structures can be designed on functional and divisional bases. Functional structures arrange distinct functions of the organization whereas divisional structures are based on location/regions, products, processes, and customer demographics. A third structure, *a matrix structure*, combines both functional and divisional structures by forming *cross- functional teams*. Human resource management is highly regulated by local and national authorities and international organizations.

Leading

Leading is a process of motivating, directing, coordinating, and communicating with teams, superiors, and stakeholders and resolving conflicts among them. Leaders motivate employees by arousing their enthusiasm and inspiring them to accomplish objectives. They direct and coordinate activities of their teams. Leaders communicate with their teams and all stakeholders to inform and influence them about the plans, progress made, problems, and expected changes. Leaders also resolve conflicts within and/or outside of their teams.

There are several types of leaders and leadership styles and there are several tools to determine them and match them for specific assignments. Extensive studies reveal relationships between personalities and jobs. Behavioral theories of leadership define leadership styles such as autocratic, democratic (i.e., participative) and laissez-faire. Grids developed on people versus output concerns of managers reveal different leadership styles for segments of concern combinations.

There is no single best leader type and leadership style that would be the best fit in general. Leadership contingency theories conclude that the best leadership style is situational.

Controlling

Controlling is observing the performance (i.e., monitoring), measuring and comparing the results to that of set objectives and producing improvements and changes as may be needed. Controlling actually starts with *planning* since what should be measured and what needs to be accomplished is established during planning. The manager, after monitoring the progress being made and with the input of observations, may determine what is working and what can be improved. A significant deviation from the set objectives may require a total overhaul of the original plans, which in turn may require a whole new organization and leader and/or leadership style.

Most Common Managerial Mistakes

A 1983 study by M. W. McCall Jr. and M. M. Lombardo concluded that the following were the top ten behaviors that prevented managers from reaching top management levels in their careers.

1. Insensitivity to others
2. Arrogance
3. Overly ambitious
4. Betrayal of trust
5. Unable to build a team or delegate
6. Specific performance issues
7. Unable to staff effectively
8. Unable to think strategically
9. Unable to adapt to superiors with different management styles
10. Overdependent on advocate or mentor

After almost four decades this list has not changed much. Although most of these mistakes can be eliminated by education and training, some that are personality related may require more than that. As long as a manager maintains an open mind on *lifelong learning* and *continuous improvement*, he/she will be more receptive to understanding and preventing mistakes.

Project Management

Obvious from the name, project management is the application of management functions to a *project* by using one or more management approaches and playing the role appropriate to the situations encountered. More explicitly, it is the application of the management process to a specific business endeavor—a project.

While there is ample evidence to indicate some form of project management has existed throughout history, the roots of modern management was introduced in the early 1900s with the Classical Approach. While the Bureaucratic Approach defined scope and organizational hierarchy, the Scientific Approach broke scope down to activities, assigned durations to them, and established a time sequence between activities. The first bar chart scheduling was introduced by Henri L. Gantt (i.e., the Gantt chart). Large projects such as the Hoover Dam (1931–1936) were built using Gantt chart schedules which was followed by Program Evaluation Review Technique (PERT) and Critical Path Method (CPM) scheduling used for major projects of medium size businesses. Through the 1980s and 1990s, the introduction of affordable personal computers and low-cost project management software allowed smaller organizations to use them to develop and improve their businesses. With the big information technology advancement of the 2000s, project management has become the means to conduct business with instantaneous online access allowing all stakeholders monitoring and participation in projects.

As the competition gets more aggressive, increasing numbers in several industries are seeking higher efficiencies and effectiveness by implementing project management to conduct their businesses. Project management techniques are tested and proven to deliver the sought for efficiencies and effectiveness if they are implemented to the correct case (i.e., the project) and if they are implemented correctly (i.e., managed correctly).

Increasing numbers show interest in project management education (i.e., one million active certified project managers as of 2019) since an increasing number of businesses seek to hire project managers. Although the increase in numbers is considered a positive outcome, there are several issues that businesses and project managers consider less than satisfactory. The main reasons for such unsatisfactory practices can be summarized as:

A. The managed endeavor is not a *project*.

- Routine tasks that should not require special techniques, resources, and practices are labelled as projects.
- *Objectives* are not defined; not defined clearly; and/or not agreed on by stakeholders.
- *Scope, cost, and time limitations* (i.e., triple constraints) are not defined; not defined clearly; and/or not agreed on by stakeholders.
- *Success* and *progress in delivering the objectives* are not defined; not defined clearly; and/or not agreed on by stakeholders.

B. The project is not appropriately *managed*.

- Wrong and/or incomplete scope definition and specification.
- Poor budgeting and scheduling; poor planning with unattainable targets and missing performance metrics.
- Poor organizing, leading, and controlling.

Currently organizations from several industries are promoting the use of project management for an excessive number of "project portfolios" that are loaded with added goals and deliverables. The ongoing desire of incorporating additional deliverables to any project management endeavor is a noble effort. Other than the quantitative ones such as the scope, cost, and time, such additional deliverables are of a qualitative nature. Every project manager should be encouraged to include additional qualitative deliverables to their projects wherever s(he) can. While a project can be considered successful even though its main—quantitative—goals have not been accomplished, such cases are rare. Stakeholders are less likely to agree to an over budget and late project delivery for the sake of *better* qualitative deliverables unless that preference is clarified up front. And if such qualitative requirements are clarified and agreed upon at the beginning, the PM should have no issues in including them in the cost and time planning of a project. However, if the opposite is true, then these projects are doomed to fail.

Construction Project Management

Due to widely differing types of construction projects, finding a template of how to best manage all types of projects may seem challenging. The basic management functions will be the same but there will be one other factor that will significantly affect how those functions are performed by the project manager. That factor is the stakeholder the project manager represents. Since stakeholders have their own and not necessarily the same goals, priorities, and contractual obligations, their managers' duties are substantially different. More specifically, while the owner's project manager will be involved in every step of every phase of a project, designer's, contractor's, and operator's project managers get involved in their respective roles and manage *their parts in the project* in line with their own goals, priorities, and contractual obligations.

Table 2–2 outlines major phases and key activities for a construction project. The owner leads as the main decision maker throughout the project whereas the other stakeholders and key participants are partially engaged. Hence, although the management functions are the same, what is managed and in what priority differs quite a bit from the standpoint of the stakeholder a project manager works for.

Cost and Time Considerations at Initial Stages of a Construction Project

As a construction project is conceptualized, the expected returns, either financial or not, are the ultimate goals. Since there will be an investment, initially the cost is scrutinized within a general target time frame. In other words, the investment budget is evaluated against a conceptualized business case. Once the project takes off to design and construction phases, the other stakeholders are expected to deliver within the time windows allocated to them.

This customary practice introduces another differing aspect for owners' and other stakeholders' project managers. The owners' project managers have to work out cost estimates before there is any detailed design and/or time schedule. They have to produce budgets that will be the decision criteria for the overall feasibility of the project.

Table 2–2 Phases and key activities in a construction project life cycle

1. Business Concept	4. Construction Contract Award

Conceptual Design
Order of Magnitude Estimate
Initial Feasibility
    Cost Budget
    Returns
Initial Checks:
    Legal
    Financial
    Other
	Mock-Up
	Designer Novation
	Design Development

2. Organizing/Selecting	**5. Construction**
Partners/Operators Delivery System Contract Pricing Designer(s) Contractor	Detailed Design Changes Contract Administration Area Handovers Substantial Completion
3. Engaging the Contractor	**6. Pre-Opening**
Bid Type Bid Documents Pre-Bid Assistance Price Confirmation Period	Functional Checks Training Calibrations
	7. Operations
	Guarantee Period Final Completion Maintenance Capital Improvements Asset Management
	8. End of Economical Life
	Renovations Sell or Repurpose Decision

On the other hand, especially the contractors' project managers who are in a position to construct the project, are in a fundamentally different position. They are in a position to build the project for a certain price (or a cost to allow for a certain profit) within a certain duration. This requires them to rework the cost estimates and align both cost and time constraints for the contracted scope.

How that can be accomplished is the subject of the following chapters in this book.

In Summary

- Managers get things done effectively and efficiently.
- The roots of modern management go back to the early 20th century when classical approaches focused on increasing production efficiency.
- The behavioral approach introduced the human element to the product efficiency process.
- Situational approaches combine parts of classical, behavioral, and other approaches with leadership styles in search of the best combination for a specific situation.

- Managers, in several roles, use their skills to manage. Managing requires decisions to be made. Ethical analysis and questioning is a vital part of the managerial decision-making process.
- Modern management process is planning, organizing, leading, and controlling in a circular, one following or pushing the other, order.
- Project management is applying modern management functions with appropriate techniques to projects. It has become the choice of conducting business in several industries since it aims to deliver project targets on time and within budget.
- Construction projects come in very different types. They have three main stakeholders and numerous key participants.
- Construction project managers' duties and priorities are quite different from each other depending on the stakeholder they represent, the delivery system, and the contract pricing type.

3 Understanding Cost and Its Elements

Introduction

Planning is the starting point in the project management process. Planning starts with breaking down the scope into activities, preparing the cost estimate for and establishing the time relationship between them. But everybody knows the close relationship between time and money. The cost of a project will highly depend on its duration, or the duration of a project depends on its cost: the resources allocated to its activities.

Unless it is specifically prioritized, project cost and duration are almost equally important for all stakeholders. Project managers may choose cost or duration as their initial criterion for planning a project. However, regardless of which constraint is chosen, the other constraint needs to be verified against the results obtained from the planning of the first one.

This chapter includes a detailed analysis of the two elements of any cost, defines perception of cost and price for buyers and sellers, and compares value against cost and price. It also presents how the elements of a cost can be used for both cost and time planning and control.

Cost Is More Than a Dollar Amount

Especially cost information on human and other resources are crucial for cost and time optimization. Once established, the schedule, budget, and other resource allocations can be used as the performance references and they can be used to monitor project progress. Also, the feedback obtained by monitoring actual progress can be used to tweak and improve performance.

The most common definition for "cost" can be simplified as it is the resources to be expended for obtaining some goods and/or services. One commonality in cost definitions is that the resources to be expended is expressed as a dollar (money) amount. In this context the words "product", "goods", and "service", as well as "money" and "dollars", are used interchangeably. In some rare cases the *cost* may be expressed in some measure other than money. One could say: "It will **cost** us 2 hours to take that route" expressing the cost in terms of time. This is especially important in cases where the duration to complete a task is more important than

DOI: 10.1201/9781003172710-3

the amount of money spent to accomplish it. Likewise, one could also say: "It will **save** us 2 hours to take that route" which not only impacts the completion duration of that task but may also reduce the expected money to be spent.

When asked the simple question, "How much does that product cost?" the expected answer is a dollar amount. For a project manager that is not a complete answer. To make that answer complete and meaningful, the PM needs the related quantity and its unit to accompany that dollar amount.

The next time you are in your favorite supermarket, take a look at how they answer that question. You will find out that they display their answers in the labels on the products they are selling. These labels provide in some big font size the dollars you have to pay but they also indicate, most probably in a much smaller font size, for what quantity of that product. While the dollar amount is unmistakably a number, the quantity and unit will differ from one product to another. Very common examples for the quantity and the associated unit are "each", "per

Figure 3–1 Price labels part a and b.

pound", "per lineal foot", "per dozen", "per ounce", "per hour", "per day", and similar others.

The indicated prices usually do not include any applicable tax in the United States. The first price tag in Figure 3–1 indicates $11.79 (plus $1.75 CRV; plus sales tax) per 35–12-ounce cans of Coke Zero. The indicated unit price of $0.337 per can does not include the sales tax and the state required CRV (i.e., California Redemption Value) and needs to be adjusted when calculating the cost.

The second price tag gives the price of $46.72 (plus sales tax) for *each* white blackout roller shade. The dimension for this product is given as 73" by 78" which can be used to calculate per foot square or per inch square costs if needed.

Elements of a Cost

Going back to the "how much does it cost?" question, the answers tell us that *cost is a rate*. It is the proportion of a *dollar amount* to a *quantity* of a product further defined by its *unit*. Figure 3–1 depicts **the elements that constitute a cost**. All elements in that cost equation, the dollar amount and the quantity and its unit, are equally important pieces of information, although the latter in most cases are mistakenly taken for granted. When the quantity and/or a standard unit for that quantity is not provided, the user will need to translate the given cost to some standard *"unit cost"* or *"unit price"* as explained further in Figure 3–2.

Managers use *both* cost elements for planning, which paves the way to organizing and controlling functions. The planning function more specifically relates to scheduling and budgeting, which establish the baseline references to evaluate project progress and take corrective action if and when needed. Since cost is a

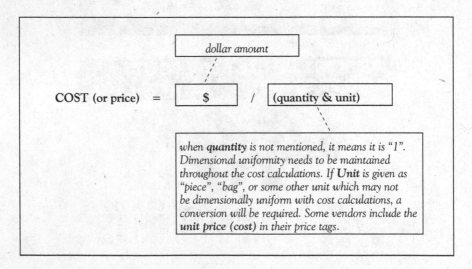

Figure 3–2 A cost has two elements: a dollar amount and a quantity with a unit.

function of both a quantity and a dollar amount, both of these variables (i.e., the elements of that cost) need to be controlled to be able to control the outcome.

For example, if the cost for bananas is given as $15.00, for some this may be quite expensive (for the year 2020). But for some other person, it could be inconclusive. If it is the cost of one banana (i.e., $15.00/piece), it could be considered to be too expensive. If it is the cost for a pound of bananas (i.e., $15.00/pound), it might still be considered too expensive for some. However, if is for a case of 10 pounds (i.e., $15.00/case), some might consider that to be a good deal. In this simple example, the elements of the cost of bananas can be defined as:

product (good) = bananas
amount to be expended = $ 15.00
quantity of goods = 1
dimension (unit) = pound (or piece or case)

In light of these definitions, the cost (or the price) for bananas could be rewritten as:

Cost of Bananas: $15.00 / pound

The activity costs and durations driven from cost data establishes the basis in preparing the project cost budget and other resources to be consumed as well as the project schedule and cash flow. The quantities driven from activity costs such as "each", "dozen", "feet", "square feet", "cubic yard", "pounds", "tons", and similar other) as well as labor and machinery/equipment hours and/or any other information such as data usage (i.e., megabytes per hour) are essential information for calculating activity durations for *scheduling* and *resource leveling*. Similar to preparing a cash flow, a *resource schedule* can be prepared by tabulating quantities of materials, labor hours, machinery, equipment, tools hours, and any other resource requirements in their activity timeline, to show their need and usage in the project schedule.

Project budget, schedule, cash flow, and resource schedule collectively constitute the basis of physical, financial, time, and resource progress planned for a project and the basis of future progress comparisons (i.e., monitoring) and making the needed adjustments (i.e., controlling). For these reasons, breaking down the cost into its elements and understanding what can be controlled and how it can be controlled for each case is essential for setting out realistic goals and success measuring metrics for a project.

Cash flow analysis may indicate the need for financing for that project. If that is the case, the related financing cost will need to be added to the cost budget as an indirect cost (i.e., overhead). In the construction industry one common practice to compensate for cashflow deficit is by owners paying a cash advance to the general contractors (G/C). In the absence of that, G/C have to address that deficit either by marking up their bid prices or by making adjustments to compensate for it (i.e., *front loading*).

Price

Cost and price are two terms which are quite easily confused. Many people use these terms interchangeably. Both terms are most commonly expressed as some dollar (money) amount but actually they do not only relate to money. In addition to money, both cost and price include additional information that can be used in planning, organizing, and controlling project tasks.

Cost and price are two words that are commonly used interchangeably. Actually, they are not the same. *Cost* is the amount of resources needed to produce a specific quantity of product (i.e., service). The cost of a product includes a variety of resources such as materials, labor, machinery, equipment, finances, technology, know-how, premises, logistics, and similar others required to realize that production but it does not include a profit. *Price* includes a profit (a margin) *added to the cost*. In other words, Price is the Cost plus Profit.

$$\text{PRICE}_{seller} = \text{COST}_{seller} + \text{Profit Margin}_{seller}$$

The cost versus price perspectives of the buyer (or user) and the seller of the product (or service) are quite different. If "x" is a dollar amount for a quantity unit of a product, for a buyer, that is the cost of that product whereas that is the price for the seller. For example, if bananas are offered at $2.50/lb. + tax, this is the buyer's *cost* for bananas, whereas, for the seller that very same $2.50/lb. is the *price*. It is important to know what perspective is taken when speaking of cost and the price.

$$\text{COST}_{buyer} = \text{PRICE}_{seller} + \text{Taxes} + \text{Other Fees \& Expenses}$$

It is also important to clearly understand what is included in the price. When buying bananas from a local market for personal consumption, the "other fees & expenses" mentioned in the previous cost equation may be ignored. The simplest example of the other fees and expenses is "shipping and handling" costs. One can see the extensive list of "other fees" in the detailed cell phone and water bills. When calculating a specific cost, all related fees and expenses need to be included.

A *mark-up* refers to the *profit margin* added to the cost and usually comes as a percentage of the cost. Profit margin is usually a percentage as well, but that percentage is based on the price and not the cost.

Unit Cost, Unit Price

Unit cost—or price—is the cost—or price—of a specific quantity of a good. Although it would be technically more correct to use *unit cost* as a reference when analyzing cost, unit price and unit cost are commonly used interchangeably without differentiating whose cost or whose price is being referenced. In this text, unit cost and unit price will also be used interchangeably for text simplification purposes.

Typically, the price for a product would be offered for a commercial quantity. The commercial standard for pipe lengths may differ between 6 to 30 feet depending on pipe type and size with a price tag for that standard length. The unit price of that pipe is its price for a specific length: If the pipe is priced for $12 for a 6' length, its unit price is $2 / foot.

In the marketplace, some goods may not be offered in the units used in cost calculations. When using unit costs, a dimensional conversion may be needed to match their units with those used in cost calculations. Failing to make that conversion is a common error that will cause dimensionally inconsistent and incorrect results.

Assembly Unit Costs

A unit price can also be prepared for an **assembly** to simplify calculations. A simple example of *assembly unit price* could be seen at the hot or cold prepared food sections of your local supermarket. In the prepared food section several hot and cold dishes are offered and they—and not the ingredients—are priced per weight or by the container size used. The prepared food in this example is the "*assembly*" of several ingredients, the prices of which are conveniently merged to one *assembly unit price*.

A similar but more complex example is drywall cost in building construction. Building dry walls, one may choose to measure and calculate how many lineal feet of studs, how many square feet of panels, how many lineal feet of joint tape, how many pieces of screws, and the amount of mudding compound that would be used. Then the related labor and other costs involved for each one of these tasks will be calculated to determine the cost of building that wall section. Alternatively, one can work out a drywall *assembly cost*. It will include the cost of all line items that are required to build a typical wall section and assign a unit price of its own (i.e., an assembly unit cost) for this type of wall. In other words, you can calculate the cost of a typical finished drywall that includes the studs, panels, screws, tape, and mudding compound for a specific dimension. That dimension can be "square feet" or "lineal foot for a specific height" depending on where it will be used.

Assembly unit costs are used to eliminate repetitive calculations and for easier use where a similar type of work is to be conducted multiple times throughout a project.

Cost Line Items

Any cost and especially an assembly cost may include costs of several items. All cost items that are included in a specific cost or the decomposed parts of a cost are referred to as *cost line items*. Table 3–1 shows the material cost line items of a cubic yard of a specific mix of reinforced concrete including its placing and curing.

Example for Unit Cost Conversions, Assembly Unit Cost, and Performance Assumptions

A company undertakes constructing buildings with typical reinforced structural foundations and floor slabs that have similar shapes and sizes. To eliminate multiple

Table 3–1 Typical reinforced concrete quantities example

Material Line Items	Quantity (per yd³)	
Concrete Mix (1 cu. yd.)	**Magnitude**	**Unit**
John's Aggregate #8 − 1½"	1800	lbs.
Smith's Sand #50 − ⅜"	1400	lbs.
Portland Type 2 Cement	480	lbs.
Fly Ash	120	lbs.
Bond enhancer admixture	0.5	gallons
Water	84	gallons
Forms	120	sq. ft.
Reinforcement #3 mild (10mm)	230	lbs.
Miscellaneous work, incl. curing	lump sum for assembly	

individual cost calculations with minimal differences, an assembly cost approach is selected. The typical material quantities are given as tabulated in Table 3–1.

Unit Cost Conversion for Materials

John's aggregate is offered for $37 per cubic yard. The mix design requires 1,800 pounds of John's aggregate #8–1¹ᐟ²" for one cubic yard of concrete. Therefore, the unit price for that aggregate needs to be converted from dollar per cubic yard to dollar per pounds:

$$\text{Aggregate Unit price} = \frac{1 \text{ yd3 aggregate}}{2,000 \text{ lbs.}} \times \frac{\$37}{1 \text{ yd3 aggregate}} = 0.0019 \frac{\$}{lbs}.$$

Using this unit price, the cost of the aggregate line item in the concrete mix will be:

$$\text{Cost of Aggregate per 1 yd}^3 \text{concrete} = 1,800 \; lbs. \; X \; 0.0019 \frac{\$}{lbs} = 34.20 \$ / yd3$$

Developing the materials list and their quantities for formwork and reinforcement steel, the materials list for this example of typical concrete takes the form as shown in Table 3–2.

After making other material unit price conversions and applying them to their respective quantities, the material cost for the example of typical reinforced concrete is calculated as depicted in Table 3–3.

Labor and M/E Unit Costs

To cast this concrete in place, several other resources, namely labor, machinery/equipment, and miscellaneous others will be required, besides materials. To calculate labor and M/E costs, both their required quantities (i.e., required durations,

Table 3–2 Material line items for typical concrete

Material Line Items	Quantity (per yd³)	
Concrete Mix (1 cu. yd.)	**Magnitude & Unit**	
John's Aggregate #8 – 1½"	1800	lbs.
Smith's Sand #50 – ⅜"	1400	lbs.
Portland Type 2 Cement	480	lbs.
Fly Ash	120	lbs.
Bond enhancer admixture	0.5	gallons
Water	84	gallons
Forms (120 sq. ft.)		
Plywood (5 uses)	3	sheets
Corner Braces	12	pieces
Pegs 1 × 2 × 12"	1	box
Wood planks 2 × 10 × 72" (10 uses)	3	pieces
Form oil	0.5	gallons
Reinforcement		
#3 (ø10mm) mild steel	230	lbs.
Rebar wire	3	lbs.
Spacers	1	box
Miscellaneous		
Plastic moisture barrier	100	sq. ft.
Burlap for cooling	120	sq. ft.
Water for wetting burlap	600	gallons

most commonly in hours) and their unit costs (i.e., hourly rates) applicable to those quantities will be required.

The hourly rates for labor is calculated by adding all direct and indirect expenses converted to an hourly basis. The indirect expenses include all taxes and fringe benefit costs such as payroll taxes, union fees, paid vacation time, insurance premiums, and similar other expenses.

Similarly, the M/E hourly rates are calculated by adding owning costs (or rental fees) and mobilization (i.e., transportation, assembly, etc.) and operating costs (i.e., fuel, electricity, lubrication, maintenance, consumable parts, operator, etc.), all converted to an hourly basis.

It should be noted that the previous two paragraphs describe the unit costs (i.e., the hourly rates) for the labor and M/E. Those hourly rates have to be applied to the hours, the quantities, these resources will be used to produce that one cubic yard of concrete.

Labor and M/E Quantities

Estimating **output rates** and/or **labor-hours** required for accomplishing some work is the biggest challenge for any estimator and/or project manager. Although

Table 3–3 Typical material costs for 1 yd³ of example concrete

Description	Take off quantity	Unit cost ($/unit)	Total cost ($/cu. yd.)
Concrete Mix			$ 143
John's Aggregate #8 – 1½"	1800 lbs.	0.019	34
Smith's Sand #50 – ⅜"	1400 lbs.	0.021	29
Portland Type 2 Cement	480 lbs.	0.13	62
Fly Ash	120 lbs.	0.1	12
Bond enhancer admixture	0.5 gallons	10	5
Water	84 gallons	0.004	0.336
Forms			$ 95
Plywood (5 uses)	3 sheets	10	30
Corner Braces	12 pieces	0.5	6
Pegs 1 × 2 × 12"	1 box	20	20
Wood planks 2 × 10 × 72" (10 uses)	3 pieces	12	36
Form oil	0.5 gallons	6	3
Reinforcement			$ 182
#3 (⌀10mm) mild steel	230 lbs.	0.75	173
Rebar wire	3 lbs.	1.25	4
Spacers	1 box	5	5
Miscellaneous			$ 36
Plastic moisture barrier	100 sq.ft.	0.1	10
Burlap for cooling	120 sq.ft.	0.2	24
Water for wetting burlap	600 gallons	0.004	2
Total Materials (rounded)			$ 456

there are several references available (i.e., RS Means Construction Cost Data), the uniqueness of projects makes it difficult to directly apply such reference data to specific work at hand. Company records and personal experience from previous similar work is very helpful, but still several judgmental adjustments may be required. Eventually after consideration the decision maker needs to make an assumption and use that assumption to complete the process.

Output Rate

Also referred to as **Hourly Output, the output rate** for labor and/or M/E is the quantity of a specific work that a resource—and/or an assembly of resources—performs during a time period.

"25 sq. ft. per hour" is an example of one painter's output for some type of wall paint job or, a specific type of an excavator's hourly output could be given as "6 cubic yards of silty clay per hour".

Resource (Labor or M/E)- Hours

This defines the time it takes that resource to produce *one unit of the end product.* It is the unit wise reversed output rate. Using the painter and backhoe examples:

- It would take *1/25 painter-hours to paint ONE square foot of wall*
- It would take 1/6 excavator-hours to excavate ONE cubic yard of silty clay.

With either output rate or resource-hours needed for one unit of the end product and the overall quantity of the same product, one can easily calculate how long it will take to complete that activity *with one single resource.* This is the quantity element of the cost of that resource.

If one carpenter can produce 80 sq. ft. of forms in one 8 hours (i.e., daily output), that carpenter can produce 1 sq. ft. of formwork in 0.1 hours.

Using similarly calculated resource-hours, the labor hours for the example of typical concrete has been calculated as tabulated in Table 3–4. Table 3–4 also provides the cost data for M/E and other resources as lumps sums.

Table 3–4 tabulates the cost elements of the line items that constitute the cost for a sample assembly cost; the cost for a cubic yard of reinforced concrete production, placement, and curing. The following sample information can be withdrawn from the data in Table 3–4:

- "Cost of John's Aggregate material is 34.20 dollars per cubic yard (i.e. $34.20 per 1800 lbs.) for this reinforced concrete design."
- "Cost of materials for this concrete mix, including concrete ingredients, forms, reinforcement, and miscellaneous other materials is 456 dollars per cubic yard."
- "The cost of this concrete mix design including it's in situ placement is 1.536 dollars per cubic yard—including materials, labor, M/E, and production overheads."

Output Rates, Resource-Hours and Activity Durations

Similar to using resource quantities in calculating activity costs, once the output rate or the resource-hour requirement per unit of activity is determined, the same resource quantity will be used in determining the duration of that activity. If a scheduler is involved, the cost estimator should be providing that quantity information to the scheduler for consistency.

Table 3–4 Sample cost elements and line items for 1 cubic yard (yd³) of typical reinforced concrete

Line Item Description	Quantity (per yd³)		Unit cost ($/unit)		Total cost $/yd³	
MATERIALS						
Concrete Mix	magnitude	unit	dollars /	unit	$	143
John's Aggregate #8 – 1½"	1800	lbs.	0.019	lbs.		34
Smith's Sand #50 – ⅜"	1400	lbs.	0.021	lbs.		29
Portland Type 2 Cement	480	lbs.	0.13	lbs.		62
Fly Ash	120	lbs.	0.1	lbs.		12
Bond enhancer admixture	0.5	gallons	10	gallon		5
Water	84	gallons	0.004	gallon		0.336
Forms					$	95
Plywood (5 uses)	3	sheets	10	sheet		30
Corner Braces	12	pieces	0.5	piece		6
Pegs 1 × 2 × 12"	1	boxes	20	box		20
Wood planks 2 × 10 × 72" (10 uses)	3	pieces	12	piece		36
Form oil	0.5	gallons	6	gallon		3
Reinforcement					$	182
#3 (ø10mm) mild steel	230	lbs.	0.75	lbs.		173
Rebar wire	3	lbs.	1.25	lbs.		4
Spacers	1	box	5	box		5
Miscellaneous					$	36
Plastic moisture barrier	100	sq.ft.	0.1	sq.ft.		10
Burlap for cooling	120	sq.ft.	0.2	sq.ft.		24
Water for wetting burlap	600	gallons	0.004	gallon		2
Materials total					$	456
LABOR						
Prepare forms in place	8	hours	35	hour		280
Cut, bend & fix rebar	3	hours	35	hour		105

C O S T   L I N E   I T E M S

ELEMENTS OF COST:
quantity (magnitude and its unit) and cost per unit of that quantity

TOTAL COST:
the product of the two elements

Table 3–4 (Continued)

Line Item Description	Quantity (per yd³)	Unit cost ($/unit)	Total cost $/yd³
COST LINE ITEMS Mix, transport, place, compact concrete; apply rough and final finishing; cover & protect; clean-up & secure area.	7 hours	35 hour	245
Curing &, maintain protections	2 hours	35 hour	70
Remove forms and clear area	2 hours	35 hour	70
Labor Total		$	770
MACHINERY & EQUIPMENT (M/E)			
Mixer Shovels Wheel barrows Placing & finishing tools Concrete vibrator Test sample kit Various hand tools	1 lump sum	110 each	110
M/E Total		$	110
OVERHEADS (O/H)			
Pickup truck Cleaning supplies Safety signs, cordoning tape Order, receive, stage materials organize and mobilize crew	1 lump sum	200 each	200
Total O/H		$	200
Total cost for 1 cu. yd. (rounded)		$	1,536

ELEMENTS OF COST: quantity (magnitude and its unit) and cost per unit of that quantity	**TOTAL COST:** the product of the two elements

In the previous example, the duration for the formwork activity can be calculated as:

$$\text{Duration }_{\text{forms}} = 120 \text{ sq. ft. x } 0.1 \text{ labor-hours/sq. ft.} = 12 \text{ labor hours.}$$

Assumed output rates that are also used in activity duration calculations will serve as the basis to monitor progress efficiency, as will be further detailed in Chapter 9.

Managing a Cost Requires Knowing Both of Its Elements

Elements of cost reveal and are essential in managing cost for the following reasons:

- Cost is a product of both elements; therefore, both quantity and unit price are equally important in estimating and controlling a cost.
- The quantities for labor and M/E resources are based on output assumptions made in the initial cost and duration planning. They need to be monitored and verified during execution of the activity. Failure to meet those output assumptions is a performance deficiency that needs to be addressed to achieve the planned cost and duration.
- Failure to identify and address performance deficiencies will cause cost and schedule overruns. Performance deficiency (or efficiency) for labor and M/E depend on several factors as detailed in Chapter 9.

Tax, Price, and Cost

Whether disclosed or not, there are several taxes that are imposed on buyers of products and users of services. When buying/selling a product/service, knowing what taxes are included and/or will be applied at the point of sale is a must for properly determining the cost.

In the United States, the State and local taxes are generally not included in the disclosed prices. Some rare exceptions to that would be the gasoline prices on marquee signs of gas stations and airline fares. Although those prices include several federal, state, and city taxes, none of that is disclosed to the customer unless it is specifically asked for. When a price tag indicates "+ tax," that applicable tax may range from 0.1% to over 20% depending on the state, county, and city where that product/service is acquired. It is always a good practice to verify if there is any applicable tax to that transaction before committing to it.

While all taxes, tariffs, dues, and all other fees embedded in the price of a good/service will be a part of the cost for the buyer/user, there are other taxes that may not be cost related. Taxes on earnings (i.e., personal and corporate income taxes, employee portion of payroll taxes, and capital gains taxes) are not costs that can be processed as expenses. The other category of taxes, the ownership taxes, can be processed as costs on a case-by-case basis. While tangible property taxes on business assets can be processed as business expenses and costs, the same does not apply to inheritance and estate taxes.

Especially when working in a new county and/or state, guidance from tax consultants is highly recommended for determining what taxes apply and which of those can be included in costs.

Can the Price Be Less Than Its Cost?

Although it is very rare, a seller may offer a price below the product's cost. As shown in our previous price equation, the price can be less than the cost if the seller chooses to apply a "negative" profit margin for that product. A seller's price, below its own cost, may or may not be a sale at loss. In some specific situations and/or as a part of a sale strategy, the seller/provider may decide to offer a product at a rate lower than its "average" cost.

Imagine that the seller put on the market 1,000 bananas which has cost the seller $1.0 each. The *price* is set at @1.50 / banana. At the end of the day, just 2 hours before closing time, the seller finds out that 900 bananas were sold for $1.50 x 900 = $1,350. At this point the seller may choose to sell the remaining 100 bananas for 75 cents each to quickly liquidate the inventory. That decision may be due to the fact that the bananas may go bad the next day, or the seller may not have proper storage and may not be willing to pay for that storage and/or the seller may want to move on to another product, say strawberries, which may have its own requirements.

Although selling the remaining 100 bananas for 75 cents each may seem to be selling at a loss, the seller will end up making a 42.5% profit from the overall sale of 1,000 bananas. If the seller insisted on selling the bananas at the $1.50 each price point, he/she would have to sell over 50 bananas to accomplish the same profit margin and then would have to deal with the left over 50.

As demonstrated in this simple example, the price and cost may differ in different operational phases. A seller may strategize to sell a certain quantity of products at one price point and then shift to a lower price for the remainder of his/her inventory. **Break Even Analysis** is a widely used tool to make such decisions.

Defining Value

Cost and price are quantitative measures. Referring to "the value" of something can be confusing, especially when that referred value can actually mean cost or price. In this book the term value will only be used as a subjective and qualitative measure. While quantitative variables can be defined with a numerical value and a unit that define that variable, qualitative variables are categorical variables that cannot be defined with magnitudes and units. Scent, beauty, impressions, emotions, flavors, satisfaction, and aesthetics are some examples of qualitative measures. Obviously, the perception of such measures varies from person to person, society to society; they are subjective variables. Some qualitative measures can be quantified to establish an objective reference. Putting a wavelength or a "temperature" number reference to a light source is an example of quantifying the color of a light which is a qualitative variable. Although it is technically possible to quantify some qualitative measures, such quantified referencing is mostly not possible and/or may not be practically meaningful for most end users.

"Building and fitting out a very high end designer casino is a challenge of its own. Working with dozens of esteemed design professionals combined with the owner, himself being the chief designer, takes that challenge to another level. All under extremely high time pressure, everything in that project had to have a full scale and fully functional mock-ups. When the mock-ups were 100% complete, the owner would inspect it in its entirety and the work could proceed only after obtaining his approval. After several reviews from a hoard of designers and the staff that would operate the casino, a mock-up prepared for one section of the gaming area was agreed to be ready for the owner's inspection. The owner found everything fine with all the finishes, colors, lighting, furniture, artwork, safety and surveillance equipment and a long list of everything else in that space but he was still hesitant to give the most needed go ahead that would keep that project on time. While all builders, designers and consultants anxiously awaited, he said, "all is fine but there is something missing here: this place does not make me feel lucky" and he left. After a couple of hours of scratching our heads to find out how a space can be designed and/or constructed to make customers feel "lucky," the owner called and gave us what we needed "remove the mirrors on the pillars and go ahead, build it."

MNH; Wynn Macau Casino Resort, 2005

One simple definition of value is, "value is what something is worth". What something is worth is very personal and subjective. The same thing may/may not be worth the same for someone else. "One person's trash is another person's treasure" is an idiomatic expression of the same.

Another definition of value is: "price is what you pay and value is what you get in return". Again, "what you get in return" in this statement is a qualitative variable. If that "return" can be quantified in the form of some "function" obtainable from the product paid for, value can be defined as the proportion of that function to cost—the amount paid for that product.

Value = Function/Cost

Since ethical behavior is driven by values and since ethical conduct is a major concern for many businesses, value has become a prominent project driver along with the triple constraints. Considering "value" from all applicable aspects is a very important effort in making managerial decisions. Because of its qualitative nature, evaluation of value may differ widely for every individual, for every organization, and for their respective personal and organizational expectations.

Value, Quality, and Cost

Although it is not uncommon to use value and quality as interchangeable terms, there are profound distinctions between these concepts. While value has its roots in beliefs that define favorable behavior, quality is a concept of perception, unless it refers to some production quality defect, which is not the subject here. Both value and quality concepts are subjective. Perfection, refinement, excellence, and completeness are some of the words that relate to quality. While both values and quality are desirable, there is no established correlation between them. Some result, service, or product may have very low perceived quality but at the same time may have very high value under the circumstances.

"More than four decades ago, I worked as a field engineer constructing irrigation and drainage channels and laying pipes in the Mesopotamian desert. The nature of the work did not allow for stationary site barracks. Everything had to be mobile to keep up with the work progress. Along with many others, fresh water was a high demand commodity and its supply was especially difficult since there were no PET bottles and bottled water available in that country at the time. Appreciating the work our teams were doing for them, the local villagers would offer us muddy water drawn from their wells. I would place that water in several glasses, lined up on my tent-office desk for hours before I could sip the water from the top without disturbing the sediment at the bottom. The water quality was way below acceptable but under the circumstances, it was a high value gesture from the locals that I gladly accepted."

MNH; Kirkuk - Adhaim Irrigation and Drainage Project, 1979.

Scholars have argued that quality and cost are directly related. While increased quality may lead to increased costs, the opposite is a harder case to make. Increasing the cost will not guarantee increasing the value and/or quality.

The direct correlation of value and quality with cost may as well be another one-way avenue which may require careful case-by-case analysis. Increasing cost may not generate and/or add value or quality. However, targeting a certain value or quality is more likely to increase cost.

Types of Values

Since value is such a broad-based qualitative and subjective concept, there may be infinite types of values for individuals and organizations. Economic, religious, political, cultural, and social values are examples of some renowned values. Beyond those, there will be organizational and business values which may or may not combine with the first set of examples. Considering the diversity of religions, cultures, societies, and political views expected to be present in a business environment, the number of possible combinations of the above-mentioned value types and the resulting variance will be staggering.

Value Priorities of Stakeholders and Key Participants

A project manager should expect that each stakeholder and key participant will have their own set of values and value priorities. It is extremely important for the project manager to identify those priorities as early as possible. Depending on locality, diversity of team members, and cultures of organizations, certain value clashes may be unavoidable. This should not hinder the success of any project. Timing for identifying and addressing possible ethical dilemmas is the key to resolving such challenges. The earlier such issues are surfaced and addressed, the less the project will suffer.

As projects progress, and especially for projects that take several months/years to complete, project managers should expect changes in the priorities of stakeholders and participants. Due to the nature of such issues, parties may or may not feel comfortable having such changes communicated through channels open to others. Private meetings in person may be a good option for checking and/or reconfirming the priorities and any changes to be expected.

Value as a Measure of Project Success

Value, as a qualitative and subjective perception, is an extremely complex concept on its own. A major project is expected to have several stakeholders, all with some type of a combination of values previously mentioned. Unless they identify their common value goals in due time and preferably at the start of the project, late changes can lead to catastrophic results. When value is introduced as a measure of success for a project, that introduced value needs to be very clearly communicated with the project management. Dealing with some vaguely defined, qualitative and subjective variable, along with the scope, time, and cost constraints, will make project success exponentially more difficult for the project manager to deliver.

Considerations for Setting Value Goals for Projects

Expecting projects to deliver some form of value is a noble target. An increasing number of businesses are introducing value in their project deliverables lists. For a project to deliver that value, that project needs to be completed within the constraints set at its start. If the stakeholders are flexible with their scope and time constraints and are willing to pay for it, there is no reason they should not pursue values as additional goals. However, as is the case for much less sophisticated deliverables, unrealistic targets fail projects. In addition, having the stakeholders' agreement on what the targeted value is can be a major challenge for project managers.

The focus of this book is on the traditional project constraints. The growing trend of introducing *value* as an additional qualitative project goal is acknowledged and appreciated. However, covering literature on such project management practices comprehensively is beyond the scope of this book.

In Summary

- Any cost in comprised of two elements; a dollar amount and an associated quantity.
- If not provided, the quantity element of a cost needs to be converted to a more conventional unit to enable the use of that cost in a dimensionally correct form.
- Price includes a profit; cost does not.
- Cost and price conception changes based on who is the seller/provider and who is the buyer/user.
- Unit cost—or unit price—is the cost/price of a good/product for one/single commonly used quantity unit.
- A cost may contain multiple line items, with each line item having their own costs.
- Value is a subjective—qualitative—variable the perception of which may widely vary among project stakeholders and key participants.
- Although becoming increasingly popular, setting value(s) as a measure of project success requires a delicate consensus among project stakeholders and key participants.

4 Cost Types and Components

Introduction

The types of costs and the general composition of a cost is the subject of this chapter. Identifying the cost types and components is essential in preparing the costs as well as analyzing and eventually controlling them. Categorization of cost types provides an indication of which costs are more likely to be controllable. Breaking a cost into its four possible components reveals what element of what component needs to be focused on for both planning and controlling purposes. The need for analyzing both cost elements in order to be able to control the outcome is reemphasized. The elements of a task's cost components contain information that constitutes the basis of resources needed to deliver that task. That information is key for cost and time planning of the project, which then serves as the project's progress evaluation basis. Details and examples are provided to explain the relationship between cost types, their components, and the elements of those components that need to be focused on in order to control the resulting cost. Since the number of activities can easily exceed several thousand for a medium size project, prioritizing which costs to focus on by categorization of their cost types is emphasized from a practicality perspective.

Types of Costs

There are four main types of costs which are direct, indirect, fixed, and variable costs. Costs can be identified with the combinations of these types, such as fixed direct, fixed indirect, variable direct and variable indirect costs.

Direct and Indirect Costs

In its simplest form, the costs of all resources that are specifically needed to produce a deliverable (i.e., accomplish an activity) are the **direct costs** of that deliverable. Of the several resources consumed in accomplishing an activity, some become a part of the end product and some do not. As long as they are specifically required for accomplishing an activity and even though they are not a solid part of the end product, all costs of such resources are direct costs of that product or deliverable.

DOI: 10.1201/9781003172710-4

To exemplify this definition consider a reinforced concrete footing. This footing is formed of concrete and rebar which are physical parts of that structural element and their respective costs are certainly direct costs. But how about the formwork, water, compacting equipment, and the various trades of work involved? They do not appear as a part of the end product but certainly they were consumed specifically for producing this footing. Therefore, they are also direct costs of this footing.

An activity may require incurring other costs of resources which are not specifically related to that activity but without those resources that activity cannot be accomplished. Such costs, which are required but not activity/deliverable specific, are **indirect costs.**

The cost of the structural design is an example of an indirect cost for that footing. If that design was only specific to that footing, that would have been a direct cost. Other indirect cost examples for this footing would be the cost of the security fence and signage around the construction site, the cost of permits, and general supervision costs.

Other examples for direct and indirect costs can be driven from the daily operations of a truck on a job site. In case that truck is only used to deliver sand for concrete production, the cost of the sand as well as that truck's cost for delivery are both direct costs for the produced concrete.

If the same transported aggregates are used partially for concrete production and partially for landscaping, the related cost is a direct cost for both concrete production and landscaping tasks and should be appropriately split between these two tasks.

If the same truck is used to deliver a variety of materials for several activities, then the owning-operating cost of the truck can be considered as an indirect cost to be distributed over several activities (i.e., a site overhead).

Unless a cost wholly or partially cannot be categorized as a direct cost then that cost is an indirect cost. If one specific indirect cost is related to more than one activity, it can be categorized as an overhead and can be disbursed to a group of activities.

If a supervisor oversees a specific activity in several locations (i.e., rebar installation for structures at different locations), then the cost of that supervision can be included in the rebar cost as an indirect cost.

However, if the same supervisor oversees rebar installation as well as other activities such as forms, scaffoldings, concrete placing, and excavation and backfill of trenches, the cost of that supervision is an overhead (i.e., indirect cost) that needs to be distributed over all of these activities.

Depending on the organization complexity, direct and indirect costs may be perceived and processed differently. It is quite common for one organization to have a head office where the general procurement research and contracting is done for several construction sites where those procured resources are used for similar activities. A good example would be a construction company that negotiates a deal with an aggregate and sand supplier to provide those materials to its various sites. Each site will incur a cost to order sand and aggregates from that supplier, to receive and store them, and process the payment when it is due. For that site, this portion of the cost is an indirect cost of its concrete cost. However, for the head office the overall site cost for concrete will be processed as a direct cost and the head office will add its own procurement process cost—as well as other head office costs—to that as an indirect cost.

Some organizations prefer calling direct costs **production costs** and indirect costs **non-production costs** as they relate costs directly or indirectly to *production* of deliverables.

Fixed and Variable Costs

Another categorization of a cost is made on it being fixed or variable depending on its usage. A cost is categorized as a **fixed cost** if that cost does not change by its usage. The monthly rent of an office is a fixed cost; that rent does not change depending on the use of that office. Likewise, the rent for a storage space is a fixed cost since its rent—its cost—does not change according to how much is stored there.

A **variable cost** changes depending on its usage. Most operating costs are variable costs whereas most owning/installation costs are fixed costs. While the rent of an office space is a fixed cost, the cost of utilities for that office is a variable cost. Another example is delivery costs; they will vary on the distance the delivery will need to be transported.

For a cost to be identified as a fixed cost, both elements of that cost, namely both the dollar amount and the quantity-unit needs to be fixed. Changes due to inflation and/or periodic changes to prices should not make related costs variable.

These two categorizations of costs are not mutually exclusive. Costs can be categorized as **fixed direct**, **fixed indirect**, **variable direct** and **variable indirect**.

Examples

Fixed Direct Costs

Costs of sand, crushed stone, cement, rebar, and water in a concrete mix are fixed costs since their quantities are fixed as per the mix design.

This statement assumes the unit prices of these ingredients do not change throughout the production period. Either cost element changing due to usage and/or timing makes the cost a variable cost. Many utility companies incentivize reducing excessive use by offering reduced unit prices when customers consume less than a predetermined quantity and/or if the consumption takes place outside daily and/or seasonal peak times. If subject costs may be impacted due to such measures, they can no longer be considered as fixed costs.

Fixed Indirect Costs

Long-term rents for site installations, storage facilities, fees for permits, and all risk insurance premiums are fixed costs which are not activity specific. The project incurs these costs and similar others like these regardless of the activities chosen to work on.

Variable Direct Costs

The power consumed by the batching plant, the fuel used by the truck mixers, and the generators that produce the power for hand tools are all direct but variable costs for a concrete mixed and poured at a job site. Although the batch plant is a site installation, it is specifically there to produce concrete for site use. Therefore, its installation cost is a fixed direct cost whereas some of its operation costs are variable direct costs for related concrete activities.

Variable Indirect Costs

Costs for supervision, quality control, safety, and security measures are examples of variable indirect costs assuming that many organizations will exceed the required minimums. Since there cannot be a maximum, such costs will vary according to preferences of the organization. Other examples for a company head office are public relations, and business development efforts. Again, the costs of such activities will vary according to preferences and aggressiveness of such efforts.

Identifying Cost Types Example

You are planning on having a dedicated business telephone. The following are the options you find at your local service provider store:

Option1. Buy an unlocked phone for $480 to be paid for in $20/month installments and subscribe to:

- Plan (a) All inclusive, unlimited data, text, and voice time for a $60 fixed monthly rate
- Plan (b) Pay as you use at a set rate for each service (i.e., data $5/Gb, $0.40/ message, and $0.25/minute for voice)

Option 2. Buy a plan that comes with:

- Plan (a) A phone and an all-inclusive fixed monthly fee of $100 for unlimited data, text, and voice
- Plan (b) A phone and a pay as you use plan at a set rate for each service (i.e., data $8/Gb, $0.50/ message, and $0.40/minute for voice)

You estimate that each month you will average 100 minutes of calls, send and receive 300 messages and will use 5 Gigabytes of data.

The following costs will be incurred at the end of a billing month (i.e., the billing cycle) during which this phone was used:

Option 1, Plan (a) total cost $ 80:

$20 phone installment: Indirect fixed cost (the phone is used to communicate with managers who are in charge of several activities)

Table 4–1 Offered phone and plan costs and cost types

Plan	Option 1 Costs		Option 2 Costs	
	Fixed	Variable	Fixed	Variable
a	$ 80 / month		$ 100 / month	
	($20 phone installment and $60 for the plan)	–	(includes phone and usage)	–
	total cost: $80/month		total cost: $ 100/month	
b	$20/month (installment)	$170/month ($25 for voice $120 for texts $25 for data)	–	$230 / month ($40 for voice $150 for text $40 for data)
	total cost: $190/month		total cost: $ 230/month	

$ 60 plan fee: Direct fixed cost for the related activities (phone was used specifically for these activities)

Option 1, Plan (b) total cost $190:

$20 phone installment—indirect fixed cost
$170—direct variable cost for related activities

Option 2, Plan (a) total cost $100; indirect fixed cost
Option 2, Plan (b) $230; indirect variable cost

In this example, the categorization of cost types may/may not have any signifi-cance if the objective is to obtain the final cost. However, identifying the cost type provides insight in determining what element of that cost is controllable.

Understanding what is variable and what is fixed for a cost enables planners to make improved decisions and choices. An example for such a case is equipment rentals. When a professional tile saw is needed for a project at home, it can be rented on an hourly and/or daily basis. That rental fee will be a fixed cost and most probably there will be daily and hourly rental options.

To make most of that rental fee, a fixed cost, the quantity of work needs to be maximized for the period of rental (i.e., lowering the unit cost by increasing the quantity against that fixed cost). One way of doing that would be laying down all the tiles that do not have to be cut first, before renting that tile saw. This way the tile saw will be used to cut tiles one after another without having to wait until the next cut and hence reduce the total rental hours which in turn will reduce the rental cost for that tile saw.

Another option would be to rent the tile saw for the whole duration of the til-ing activity—say a day—and cut the tiles as they are needed in the installation process. If the durations are estimated for both of these options, a decision based on the comparison of their costs would be made. This cost planning also establishes what needs to be controlled in order to stay within that targeted cost.

Such alternatives would not be considered if the project involved a quantity where rather than renting that tile saw, buying one would turn out to be more feasible. In that case, the cost of the tile saw would be a fixed direct cost to be added to that tile activity.

Tables 4–2 and 4–3 provide cost type categorizations for common construction job site and construction company head office costs.

Knowing what costs and what part of those costs are variable will enable managers to control those parts in their cost control efforts. Preventing excessive use and/or wastage, and timing for better rates are some measures the managers can take to lower task costs.

An example for such a case is electrical power use. Almost all power companies offer increased unit prices after exceeding a pre-set total use quantity and/ or for usage at peak hours of demand. Cutting off power for chargers when they are not charging tools and equipment will reduce the overall power consumption quantity. In addition, charging tools and equipment can be done when the power

Table 4–2 Cost type examples for a construction site

| | Type of Cost | | | |
| | Direct | | Indirect | |
Examples of Costs	Fixed	Variable	Fixed	Variable
Assets on Job Site[1]: Plant, equipment, vehicles, hardware, software owning costs and/or long-term rentals, site installations and improvement costs; office, warehouse, stockyard owning costs, and/or long-term rentals; business permits, licenses, registration fees, as they can be associated directly or indirectly with productions	✓		✓	
Operating and maintenance costs for all assets on site[1]		✓		✓
All supervisory and support personnel salaries and benefits			✓	
All production labor costs		✓		
Head Office overhead distributed to job sites				✓

[1] These costs can be considered direct or indirect; they are indirect costs for site activities but they are direct project costs for the company head office.

Table 4–3 Cost type examples for a construction company head office

| Examples of Costs | Type of Cost | | | |
| | Direct | | Indirect | |
	Fixed	Variable	Fixed	Variable
Assets and Premises Related Cost Examples [1]				
Asset replacement/reserve fund allocations; Furniture—Fixture-Equipment fund allocations; plant, equipment, vehicles, hardware, software owning costs and/or long-term rentals, site installations and improvement costs; office, warehouse, stockyard owning costs and/or long-term rentals; business permits, licenses, registration fees	✓			
Operating Cost Examples				
Utilities (electricity, gas, fuel, water, internet service, telephone service, wifi network, etc.), operating supplies and consumables, sales and marketing promotions, travel and lodging expenses, guarantees, warranties, bonds, insurances, legal fees related to general operations			✓	
Payroll/Labor Cost Examples				
a) Human resources, finance, accounting, sales and marketing, and other permanent employees of the headquarters.	✓			
b) Salaries and fringe benefits of employees who work on the job (production) site and who are hired on a yearly basis.	✓			
c) Payments for hourly hired workforce at the job site.		✓		
Costs incurred for malpractice indemnifications, penalties, damages, legal fees, and consultant fees for sporadic cases				✓
Periodic dues and contributions to professional asssociations, continued education and training for employees, publications, public relations expenses [2]			✓	

[1] Depreciation is excluded and may not represent the fund allocations mentioned here. See section "Depreciation Is Not a Cost" for more details on depreciation.

[2] Some of these expenses can also be categorized as variable costs.

companies offer lower rates during off peak hours which will decrease the power cost for the subject tools and equipment.

Identifying indirect costs and their origins is critical in order to keep those costs under control. Costs cannot be controlled if there will be a charge from an unexpected source and for an unknown amount beyond the management's control. In the case of a construction site in the downtown area of a major city, it would be reasonable to expect some traffic violation tickets. However, for a high-voltage transmission pillar erection site, if the site gets a ticket from the forest stewardship on the basis that a certain permit was not obtained by the head office, this is a case of an indirect cost beyond the project manager's control. Another example is the head office overhead distribution to projects. This is commonly an indirect cost projects incur and those costs are beyond project managers' control.

Components of a Cost

Any cost can be broken into all or a partial combination of the following four main components:

- Material cost
- Labor cost
- Machinery, equipment, and other direct costs (M/E)
- Overhead (O/H)

Of these four components, material, labor, and machinery/equipment costs are direct (fixed and/or variable) costs. However, the overhead component of a cost differs from the other three components. Overhead costs are indirect costs and can include all other three cost component types that are indirectly incurred.

Material Component

As the name implies the *material cost* is the cost for the materials used in delivering the product and/or service. The material costs are direct costs. The cost of the product includes the losses (wastage) expected in the process and attic stock (stored reserve for future use) as may be required. Wastage is a direct cost, being a direct result of production. However, if some materials are wasted due to improper storage and handling and not as a result of production, that wastage is an indirect material cost. Attic stock is an indirect cost. Depending on project stakeholders' preferences and/or as may be required by project specifications, there may be other costs, directly or indirectly related to material costs. Additional testing and certification beyond that of the manufacturers' regular testing and certification is an example of such additional and indirect costs.

Labor Component

The labor component includes all *direct labor costs* including the fringe benefits. If the process involves partial supervision and specific tasks like quality control, those labor costs should be distributed to the related tasks and should be included in those tasks' labor components as indirect costs.

Some construction activities may not allow continuous work for crews. Concrete and painting are examples of two activities that mandate wait times in between pours and coats. For such activities, the manager has several options to consider:

a. Let crews wait and charge their cost—including wait times—to the labor component of that activity.
b. Let crews wait and charge their cost—including wait times—as an overhead cost.
c. During the wait times, engage crews in some other activities and distribute their cost to the labor cost components of all those activities as well as the main activity.
d. Outsource (i.e., subcontract) the activity and charge the cost to the labor component of the activity.
e. Find a method and/or additional and/or different equipment to minimize wait times.

On a project site a concrete crew is assembled to place and finish concrete at several areas. If that project can keep this crew busy full time, the entire labor cost of that crew is a direct cost for placing–finishing concrete. If there are waiting times in between concrete pours, those idle times of this crew is an indirect labor cost for that concrete activity. If the same concrete crew can carry out some other work while they wait for the next pour clearance, like cleaning and preparing plywood for forms, the related labor can be added to "forms" task as a direct cost. Likewise, supervision for this crew can be either all direct or partially direct depending on the supervisors' time being totally or partially dedicated to a specific activity or not.

Many projects may not have the volume of works such as the concrete example to justify establishing a specific crew for them. A reinforced concrete building structure is such a case where subcontracting discontinuous work should be seriously considered. Each floor's concrete pour will require a wait time normally ranging from 3 to 28 days before the next pour above can take place. A subcontractor undertaking setting formwork, rebar installation, and placing concrete work at different job sites (i.e., projects) may have its crews work continuously making use of those waiting times of similar projects. The subject subcontracting cost should be compared to all direct and indirect costs when such decisions are to be made.

Another option is finding another construction method and/or equipment that would reduce those wait times and therefore reduce the relevant labor cost. Using faster setting and faster strengthening concrete, using higher strength concrete, using prefabricated elements, increasing the number of overall forms and scaffolding sets to allow keeping them in place to support the second and third consequent pours above are some of the examples of options to be considered. Naturally, the costs of all of these options are to be compared before making the final decision.

Machinery, Equipment, and Other Direct Costs (M/E) Component

The third cost component is the *direct costs of machinery and equipment and any other tool, software and/or hardware* needed specifically to produce the needed good or service. From the receiver's perspective, the costs related to the batching plant that mixes the concrete, the truck mixer that transports that concrete to the site, the pump used to pump that concrete to where it will be poured, and costs related to all other tools, machinery, and equipment used in the production process are direct costs in this cost component. The cost of maintaining a workshop at that job site that services other machinery and equipment as well as those previously mentioned will be either an indirect cost for the overall machinery and equipment cost of the entire site or, the portion of that overall workshop cost apportioned specifically for the concrete equipment will be a direct cost for producing and placing concrete. Other premises/utilities/services like the heating–cooling, lighting, ventilation, internet service, and staging areas that are not needed specifically for the production and delivery are examples of *indirect* machinery and equipment costs that will more appropriately be included in the *overhead component*.

Overhead (O/H) Component

The overhead(O/H) cost component is the only component that covers all indirect fixed and variable costs attributable to that product or service. O/H by definition is an indirect cost.

When setting up shop for a business, the investment is done for a series of products including some specialty items that could be the centerpiece for that business. The investment will include some equipment, fixtures, and improvements to the shop which will be *depreciated* or, in other words, be recovered in time through tax allowances. Such investments that are subject to depreciation should not be included in any cost components. The operating costs such as rent, utilities, sale promotions, license fees, and similar others are needed for a business to produce that good and/or service, but they are not needed for producing that specific good and/or provision of that specific service. These are *general* costs that a business needs to incur to run as a business, but they are not specifically required to produce one specific product. The costs that cannot be directly linked to a specific good/service are *overheads (O/H)* and they are **distributed over the goods/services** as a separate component.

Table 4–4 provides an estimate prepared for a cubic yard of reinforced, slab on grade concrete to be put in place at a generic location. This estimate clearly indicates the line items and their elements for the cost components of this activity. This table depicts the quantity and the unit price elements of the cost components for the project manager's evaluation and comparison for alternatives such as delivered ready mix concrete. In addition, the project manager can identify which elements of those line items he/she needs to focus to control the overall cost. While it is most likely the unit prices will be pretty much fixed in the case of the direct

Table 4–4 Typical cost component details for 1 cubic yard of reinforced concrete

Component	Line Item Description	Quantity (per yd³)		Unit cost ($/unit)		total cost $/yd³
	MATERIALS	magnitude	unit	dollars /	unit	
	Concrete Mix					$ 143
	John's Aggregate #8 − 1½"	1800	lbs.	0.019	lbs.	34
	Smith's Sand #50 − ⅜"	1400	lbs.	0.021	lbs.	29
	Portland Type 2 Cement	480	lbs.	0.13	lbs.	62
	Fly Ash	120	lbs.	0.1	lbs.	12
	Bond enhancer admixture	0.5	gallons	10	gallon	5
	Water	84	gallons	0.004	gallon	0.336
	Forms					$ 95
	Plywood (5 uses)	3	sheets	10	sheet	30
	Corner Braces	12	pieces	0.5	piece	6
	Pegs 1×2×12"	1	boxes	20	box	20
MATERIALS	Wood planks 2×10×72" (10 uses)	3	pieces	12	piece	36
	Form oil	0.5	gallons	6	gallon	3
	Reinforcement					$ 182
	#3 (ø10mm) mild steel	230	lbs.	0.75	lbs.	173
	Rebar wire	3	lbs.	1.25	lbs.	4
	Spacers	1	box	5	box	5
	Miscellaneous					$ 36
	Plastic moisture barrier	100	sq.ft.	0.1	sq.ft.	10
	Burlap for cooling	120	sq.ft.	0.2	sq.ft.	24
	Water for wetting burlap	600	gallons	0.004	gallon	2
	Materials total					$ 457
	LABOR					
	Prepare forms in place	8	hours	35	hour	280
	Cut, bend, and fix rebar	3	hours	35	hour	105
	Mix, transport, place, compact concrete;	7	hours	35	hour	245
LABOR	apply rough and final finishing; cover and protect; clean-up and secure area.					
	Curing and maintain protections	2	hours	35	hour	70
	Remove forms and clear area	2	hours	35	hour	70
	Labor Total					$ 770

(Continued)

Table 4–4 (Continued)

Component	Line Item Description	Quantity (per yd³)	Unit cost ($/unit)	total cost $/yd³
	MATERIALS	magnitude unit	dollars / unit	
	MACHINERY & EQUIPMENT (M/E)			
M/E	Mixer Shovels Wheel barrows Placing and finishing tools Concrete vibrator Test sample kit Various hand tools	1 lump sum	110 each	110
	M/E Total			$ 110
	OVERHEADS (O/H)			
O/H	Pickup truck Cleaning supplies Safety signs, cordoning tape Order, receive, stage materials Organize and mobilize crew	1 lump sum	200 each	200
	Total O/H			$ 200
	Total cost for 1 cu. yd. (rounded)			$ 1,536

costs (i.e., material, labor, and M/E costs), the quantities and durations used are more likely to determine the actual final cost of the project. For the indirect (i.e., O/H) costs, each line item may be scrutinized if the project manager believes some savings on this component will also help.

Depreciation Is Not a Cost

Depreciation is not a *cost*; it is an allowance to recuperate the initial cost of an asset over that asset's useful life. It is an allowance of spreading the initial investment amount of an asset over a period of time, at the end of which the business may consider replacing that asset. When businesses start, tax codes do not allow them to include their entire initial investment expenses in their first-year tax returns if the expected useful life of those assets are longer than a single year. Instead they are allowed to gradually expense their initial investments by depreciating them.

Assets such as machinery, equipment, and technology hardware and software reduces in value over time due to use, age, or obsolescence. Businesses are allowed to reduce their taxable revenues by a depreciation amount as if that depreciation were actually expensed. That compensates for lack of taxation relief due to not expensing the initial investment cost in the first year. Using the depreciation allowance, businesses not only recover their initial investment costs but also accumulate funds to replace those assets when those assets expire their useful lives. *Reserve Funds* and *Furniture, Fixture,* and *Equipment (FFE) Funds* are examples of how businesses accumulate asset replacement funds.

Depreciation is a *non-cash expense.* It is a good incentive for businesses to recover their initial investments and build funds to replace them at the end of their useful lives and/or when needed. It will be a mistake—and an overcharge—if the entire investment cost of business assets were to be included in either M/E or O/H components. Instead, by using an appropriate depreciation method, usage based proportional distribution of such costs to projects is recommended. Straight Line, Output, Working Hours, Mileage, and Depletion methods are examples of many depreciation methods that can be used to determine such costs before proportioning them to a specific project.

What Do Cost Types and Components Indicate?

A PM can extract valuable data by identifying the cost types and segregating it to its components. That data can then be used in cost and time planning of activities and can establish the basis for monitoring and if needed, correcting those plans (i.e., controlling).

Cost types indicate the drivers of a cost. If it is a fixed cost, it can be minimized with maximizing the usage. If the cost is a variable one, it indicates usage will increase or decrease that cost. While a direct cost indicates it is the activity itself that drives that cost, that cost focus needs to be elsewhere in the case of indirect costs for that same activity.

Naturally the combinations of cost types provide combined indicators which can be summarized as:

- Fixed Direct Costs: Maximize usage; focus on activity itself
- Variable Direct Costs: Focus on usage/waste/efficiency for activity itself
- Fixed Indirect Costs: Maximize usage on all related activities
- Variable Indirect Costs: Focus on usage/waste/efficiency for all related activities

Breaking the cost of an activity into its components also reveals valuable planning and control information about that activity. With this breakdown, the composition of that cost, with its take-off quantities, estimated unit and final costs are revealed, which are the basis for the targeted performance efficiencies, from which the activity durations are calculated. During the execution, this information will direct the PM to the essence of what needs to be addressed for making improvements.

In Summary

- There are four types of costs.
- Identifying the type of a cost reveals important information for planning and controlling that cost.

- Any cost contains at least one or all of material, labor, machinery/equipment, and overhead components.
- Planning and controlling costs requires analysis of all elements of all cost components.
- Cost type indicates a PM's controlling limits on that cost.
- Knowing what can be controlled and to what degree it can be controlled helps project managers in selecting and prioritizing most controllable activities.

5 Structuring Project Cost

Introduction

"Divide each difficulty into as many parts as is feasible and necessary to resolve it".

René Descartes 1596 ~1650

Planning initiates the process of managing a project. Project planning comprises cost and time planning, which start with establishing what needs to be done to deliver a project. A project scope covers everything that a project delivery requires. Regrouping the previous three statements, the very first step in managing a project is determining the scope, what needs to be done to deliver the main goal of that project. This chapter demonstrates how the scope can be broken down to activities, to a detailed roadmap to project goals. The work breakdown structure (WBS) is that roadmap. It reveals the structure of tasks that need to be completed to deliver the project goals. WBS tasks loaded with appropriate resources is introduced as the Project Cost Breakdown (PCB). The advantages of scope and cost sharing the same breakdown structure is explained. Examples on WBS demonstrate how required detail levels may vary at different management levels and from each stakeholder's priority perspective. The chapter also covers creating a Work Package Dictionary and how that can be used to serve as the basis for generating resource totals for the project. Extensive examples that represent different views of stakeholders' project scope and their WBS detail levels are included.

Project Planning Process

The initial step in managing a project is planning. Planning is establishing the goals of the entire project and determining how those goals can be accomplished. More specifically, planning includes cost and time planning of the project scope so that the basic scope, cost, and time goals of the project can be achieved. The scope, cost, and duration, triple constraints of a project, are interactive; if one changes, the other two change accordingly. It is the manager's job to determine and/or verify if and how the project goals can be delivered within all of those triple constraints. For a specific scope, the manager has to determine the corresponding

DOI: 10.1201/9781003172710-5

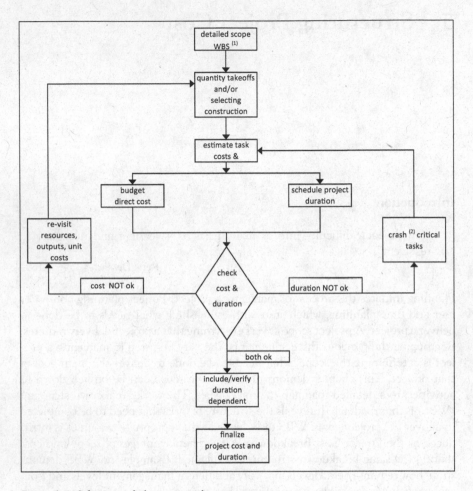

Figure 5–1 Schematic of the project planning process.

(1) WBS: *Work Breakdown Structure*
(2) Crash: *Shorten task duartion by adding resources; might change critical path.*

cost and the duration congruently since these two constraints will interactively change. In other words, for a specific scope, the cost and the time it will take to complete that scope at that cost will be calculated simultaneously as the first step in planning that project. While the summation of the direct costs of tasks will provide the project's direct cost budget, the activity durations calculated as a part of cost estimating need to be scheduled with using the logical sequence of the chosen construction method so that the resulting project duration can be verified to meet the project's time constraint. Once the project duration meets the time constraint, that duration's dependent indirect costs (i.e., overheads) are to be calculated and added to the direct costs to finalize the total project cost budget and hopefully fall within the cost constraint.

This process assumes a fixed project scope and works toward cost wise and duration wise acceptable plans for that scope. Preparing a well-defined scope and using it for both cost and time planning not only eases the process of going back and forth between cost and time plans (i.e., budget and schedule) but also makes it possible to track the nature and the magnitude of *scope changes that should be expected as the project progresses.*

For these reasons, preparing a Work Breakdown Structure (WBS) is the essential first step in project planning. Establishing the WBS and using it consistently for cost and time planning provides a structured approach to monitor interactive changes between cost and time variables. It makes it easier to deal with the three-way interactive triple constraints. With a detailed and structured scope, monitoring the remaining two interactive variables, cost and time interacting with each other, is a less complicated process as long as that scope does not change. Naturally the scope can change during a project, but when that happens it is much easier to track and demonstrate the impact of that scope change on project cost and duration.

Initiator of the Project Planning Process: Scope

Scope of a project, or as commonly referred to as *scope of works* in construction projects, is the entirety of what needs to be done to accomplish project goals. Accomplishing those goals usually requires the accomplishment of several smaller parts of the final project deliverable. In other words, the ultimate project goal comprises several smaller deliverables that collectively constitute that ultimate goal.

If the goal of a project is developing a fully operational hotel property, that ultimate target will be made up of several parts which will include—but not be limited to—financing, design, and construction from several disciplines, equipment, and furnishings. Every activity required to put together that hotel property is a part of the scope of works of that project.

Scope of works may include some products and/or services which may not be easily identifiable as a part of the end product/service. While the design services of an architect or any other designer may be prominently obvious for all, the legal and tax consultants' vital inputs for the project will be hard to be noticed by everyone. Such services are examples of essential activities that enable the delivery of project scope of works. Scope of works also includes temporary and/or some other work that does not become a part of the final product but are required for constructing the permanent works. Formwork and scaffolding for in situ concrete are examples of temporary work that is required to construct permanent parts of a project but that actually do not become a part of the final product.

Scope of works for a project depends on the method/technique chosen to carry out and deliver sections of that project. A medium rise building's structure can be designed and constructed as a wood, masonry, or reinforced concrete structure. Each approach will introduce its own challenges, require different trades, tools, and equipment, and their own activities that require a sequencing of their own. In some cases, the volume of work dictates the method to be used, requiring a quantity takeoff to take place before anything else. It will be the total amount of excavation that would either direct the decision maker to use manual labor or heavy equipment to carry out that work. Furthermore, the method /technique utilized to perform some sub-parts of the scope governs the timing/sequencing relationships (i.e., finish to start, start to start, etc.) between activities.

Establishing a structured scope definition presents several advantages later on during the project when cost, budget, and time schedule revisions have to be made for various reasons. Determining what sub-parts (i.e., activities) need to be carried out in order to accomplish one or more sections, or the entirety of the project, is the hierarchical definition of project scope. WBS provides that hierarchical order. It enables cost and time progress updates on the same basis, tracking of cost and time impacts both due to encountered changes to the initial scope and due to cost and time variances during project execution.

Detailing Scope by Breaking It Down to Tasks: Work Breakdown Structure (WBS)

Preparing the Work Breakdown Structure (WBS) is generally the first step in project planning. It details the scope, what needs to be done, so that the answers to "how", "for how much", and "when" can be prepared. The project scope for construction projects is defined in multiple documents, complementing and supplementing each other. Examples of these documents are the Invitation Letter/Bid Announcement, General Conditions of Contract, Special Conditions of Contract, Scope Statement, Design Intent Statement, Drawings, Specifications, Benchmarks, and similar other documents which are included in the bid process and/or later become a part of the contract between the stakeholders.

The detailing of scope by breaking it into smaller, better defined, and individually assignable and manageable packages, the WBS identifies what needs to be done to accomplish the goals of the project. Once it is established and the tasks are assigned costs and durations, the WBS provides a detailed roadmap for accomplishing project scope. Based on the assumed scope, it provides what cost and duration is to be expected for the project. Summation of WBS task costs constitutes the basis of the project cost budget and sequencing the tasks' assigned durations from the project schedule.

The lowest level to which the work is broken down is called the **work package or task**. The number of levels needed to define that work package depends on how *assignable/manageable* that specific task is. At any level, if some work has its scope clearly defined with its required resources, duration, and deadline as well as progress measuring metrics and if it can be assigned independently to a manager, then that level of work is a work package and no further breaking down is needed. In that respect, the very same WBS of a project may be used in different breakdown levels by different management levels; the top managers using the top WBS levels and the team leaders using the details at task levels.

All work defined between the project level and the work packages level are named as the *elements of WBS*. The branches of the highest level may have different levels of elements before reaching to work packages in that line of decomposition of related work.

WBS Should Cover the Entire Scope

Regardless of how simple a project can be, it is not a one-step process. A project may include work from several different resources. Breaking down the overall project goal into smaller parts helps define those parts with higher clarity and makes them assignable to and manageable by team members. The most important things at this point are to make sure:

- Nothing is left out and
- The detail level is not overwhelming

Sometimes referred to as the 100% Rule, wherever some section of work is decomposed to its smaller parts, the total scope of those smaller parts should add up to the total work scope of that section. If there are some known exclusions, they should be explicitly noted and either accounted for with some appropriate measure such as a contingency or left out with the proper explanation/notification. As will be explained in detail in Chapter 7, contingency is not the only measure a manager can use to account for exclusions. The work package level detail needs to provide sufficient but not overwhelming information. Overdetailing may adversely impact the implementation of cost and time management procedures.

Example: Dinner Party Project

A special dinner party is outlined next. This project will be used to demonstrate how this concise scope definition can be elaborated on by using a WBS.

Project: A backyard dinner party for the upcoming special (date specific) occasion.
Outline: 24 ~30 close friends and family for cocktails and appetizers followed by a sit-down three-course dinner with live music in the backyard. The target cost is $4,000.

The first step in planning this party is expanding the provided concise scope to serve for the purpose of this project. It will start with preparing a list of things that are expected. There would be some research, procurement and engagement activities can be grouped under Logistics, some music and fun can be grouped under Entertainment and naturally some Food and Beverages. Table 5–1 provides the WBS in line with that grouping of activities for this project; a list of activities grouped under these titles. What is included in Table 5–1 is a *wish list* WBS at this point. The cost and availabilities for our specific date need to be verified before proceeding any further.

Table 5–1 A Work Breakdown Structure example for an elaborate and special dinner party

DINNER PARTY

1. LOGISTICS

1.1. Tables and Chairs
1.2. Tabletop selection
  1.2.1. China
  1.2.2. Silverware
  1.2.3. Glassware
  1.2.4. Table cloths
  1.2.5. Napkins
1.3. Name cards, candles, flowers
1.4. Kitchen and bar setup
1.5. Garbage cans
1.6. Invitation
  1.6.1. Make list
  1.6.2. Email & confirm
1.7. Select & engage
  1.7.1. Entertainer
  1.7.2. Caterer
  1.7.3. Servers, barperson
  1.7.4. Photographer

2. ENTERTAINMENT

2.1. Set up stage and dance floor
2.2. Sound system and lights
2.3. Set up projection screen

3. FOOD AND BEVERAGE

3.1. Food
  3.1.1. From the caterer
    3.1.1.1. Appetizers
    3.1.1.2. Salad and dressings
    3.1.1.3. Main course (3 options)
      3.1.1.3.1. Steak
      3.1.1.3.2. Fish
      3.1.1.3.3. Vegetarian
  3.1.2. From the host
    3.1.2.1. Cheeses and crackers
    3.1.2.2. Banana bread
3.2. Beverages
  3.2.1. Wines
    3.2.1.1. Reds
    3.2.1.2. Whites
    3.2.1.3. Bubbly
  3.2.2. Liquor & Cocktails
  3.2.3. Soda, water, coffee

Table 5–2 Demonstration of WBS levels for the Dinner Party Project

Level	Selected Elements of Dinner Party WBS (from Table 5–1)
1	(0) Birthday Party Project
2	1. Logistics
2	2. Entertainment
2	3. Food and Beverage
3	3.1 Food
4	3.1.1. From the caterer
5	3.1.1.3. Main course options
6	3.1.1.3.1. Steak

In this example WBS, the project has been subdivided into three main work groups and each group has multiple sub-groups. Each group and sub-group is assigned numbers that indicate their grouping and their level of detail. The highest level of work is the project and it has the *lowest detail level*. The objective of a WBS is to decompose the highest level activity to lower level activities which are more specifically defined.

Table 5–2 demonstrates the six levels before reaching a work package in the Food and Beverage branch of the given WBS and one of the work packages at that level is "Steak". However, in the same WBS (Table 5–2), the Entertainment branch goes only to level 3 before reaching the work packages in that branch.

Compiling Detailed Task Information: Work Package Dictionary (WPD)

The **Work Package Dictionary** is a document that provides all relevant information for all work packages of the WBS. It is the ultimate task reference book. The WPD is expected to provide all task information that is practically possible to include; such as:

- Scope specifics
- Duration and schedule (i.e., start and end dates)
- Milestones to meet
- Expected cost and selling/contract price
- Other resources to be used
- Assumptions of the estimator and scheduler
- Progress measuring metrics
- The person in charge
- Specific exclusions
- Remarks
- Any other relevant information

A typical WPD form is provided in Appendix B.

The WBS supported by its WPD constitutes the entirety of available project information at the time of preparation. It is a key document providing detailed information on the triple constraints. While the scope details can be used to verify the intended project goals coverage, the cost data will be used for financial planning and cost control purposes as well as progress payments. There will be forward and backward feedback between the WPD, the project schedule, and the project cost estimate.

The WPD is a compilation of activity data, especially detailed cost data and durations either assigned or driven from the cost data. Using WPD data for both budgeting and scheduling establishes a much needed consistency in planning them. Once the task costs are determined, the project duration will be obtained by scheduling these tasks with their durations assigned to them when calculating their costs. If the schedule does not satisfy the project time constraint, the durations of critical activities can be shortened (i.e., crashed) by introducing additional resources. For shortened task durations, additional resources and/or alternative construction methods may be required that may increase the initial cost estimates. Similarly, lag times (i.e., task slacks) information from the schedule may result in alternative considerations for estimating cost for those tasks. Crashing requires that the planner keep an eye on the project critical path. When the critical path activity durations are shortened (i.e., crashed), it is quite possible that the critical path may shift to some other path. The fact to be remembered is that only crashing the critical path activities will shorten the project duration by allocating additional resources to those activities. Crashing the activities which are not on the critical path will increase the project cost but it will not decrease project duration.

This back and forth between budgeting and scheduling will be repeated until satisfactory results are obtained for both constraints. This process requires updating the WPD with the new task information at each iteration. Since each such iteration will have its own assumptions for estimating costs and durations, updating the WPD with those assumptions is critical for keeping track of those assumptions and their results. Using the same structure and activities makes this process easier and more consistent. The same ease and consistency will be extended to cash flow, resource schedule, and progress payment schedule preparations and applications.

Structured Project Cost: Project Cost Breakdown (PCB)

Every work package of the WBS is expected to be assigned a cost. These costs can be given, guessed to the best knowledge at the time (i.e., guesstimate), or estimated. Depending on the size of the project and/or depending on the magnitude of the task's cost, that cost can be further detailed by revealing its components. Furthermore, some of the cost line items of any component may be so significant that it may require a further breakdown of its own. Unless there is a specific reason

Table 5-3 Project Cost Breakdown (PCB) for Dinner Party Project in organization chart format

DINNER PARTY ($4,500)

1. LOGISTICS ($1,200)

1.1. Tables and Chairs (caterer $ 300)
1.2. Tabletop selection (caterer $ 600)
   1.2.1. China
   1.2.2. Silverware
   1.2.3. Glassware
   1.2.4. Table cloths
   1.2.5. Napkins
1.3. Name cards, candles, flowers ($100)
1.4. Kitchen and bar setup (caterer & bar staff)
1.5. Garbage cans (server)
1.6. Invitations (self)
   1.6.1. Make list
   1.6.2. Email & confirm
1.7. Select & engage ($ 200)
   1.7.1. Entertainer (incl. in section 2)
   1.7.2. Caterer (incl. in section 3)
   1.7.3. Servers, barperson ($ 200)
   1.7.4. Photographer (good friend, no fee)

2. ENTERTAINMENT ($ 500)

2.1. Set up stage and dance floor
2.2. Sound system and lights
2.3. Set up projection screen

3. FOOD AND BEVERAGE ($ 2,800)

3.1. Food ($ 2,300)
   3.1.1. From the caterer ($2,100)
      3.1.1.1. Appetizers ($ 400)
      3.1.1.2. Salad and dressings ($ 200)
      3.1.1.3. Main course ($ 1,500)
         3.1.1.3.1. Steak
         3.1.1.3.2. Fish
         3.1.1.3.3. Vegetarian
   3.1.2. From the host ($ 200)
      3.1.2.1. Cheeses and crackers
      3.1.2.2. Banana bread
3.2. Beverages ($ 500)
   3.2.1. Wines
      3.2.1.1. Reds
      3.2.1.2. Whites
      3.2.1.3. Bubbly
   3.2.2. Liquor & Cocktails
   3.2.3. Soda, water, coffee

Note: No dollar amount for a task indicates its cost is included in the upper level task.

Table 5-4 List format PCB for Dinner Party Project with notes

Dinner Party			notes
1. LOGISTICS		$ 4,600	
		– $ 1,100	
1.1. Tables and Chairs		$ 300	1.1 & 1.2 rental prices by the caterer. Includes setting up but excludes placing them in crates after the party. Breakage for china and glasses will be separately charged.
1.2. Tabletop selection		$ 700	
1.2.1. China	120		
1.2.2. Silverware	120		
1.2.3. Glassware	120		
1.2.4. Table cloths	120		
1.2.5. Napkins	120		
1.2.6. Breakage contingency	100		
1.3. Name cards, candles, flowers		$ 100	
1.4. Kitchen and bar setup		incl.[1]	Self, servers, and barperson will do the work; no extra fee foreseen.
1.5. garbage cans		incl.[1]	
1.6. Invitation		incl.[1]	No charge for "self" time.
1.6.1. Make list	incl.[1]		
1.6.2. Email and confirm	incl.[1]		
1.7. Select and engage		incl.[1]	No charge for self time; the respective fees for services included elsewhere.
1.7.1. Entertainer	incl.[1]		
1.7.2. Caterer	incl.[1]		
1.7.3. Servers, barperson	incl.[1]		
1.7.4. Good Friend Photographer	incl.[1]		

(Continued)

Table 5-4 (Continued)

Item	Amount	Amount	Amount	Note
2. ENTERTAINMENT			$ 500	
2.1. Entertainer		$ 500		
2.1.1. Set up stage and dance floor	incl.(1)			Included in entertainer's fee.
2.1.2. Sound system and lights	incl.(1)			
2.1.3. Set up projection screen	incl.(1)			
3. FOOD ANAD BEVERAGE			$3,000	
3.1. Food		$ 2,300		
3.1.1. From the caterer	2,100			
3.1.1.1. Appetizers				
3.1.1.2. Salad and dressings				
3.1.1.3. Main course (3 options) 1				Caterer price covers all 3 options but does not provide further details.
3.1.1.3.1. Steak	incl.(1)			
3.1.1.3.2. Fish	incl.(1)			
3.1.1.3.3. Vegetarian	incl.(1)			
3.1.2. From the host	200			
3.1.2.1. Cheeses and crackers				Guestimate numbers to be verified.
3.1.2.2. Banana bread				
3.2. Beverages ($500)		$ 500		
3.2.1. Wines	500			
3.2.1.1. Reds				Safe numbers; some guests will bring their own as well.
3.2.1.2. Whites				
3.2.1.3. Bubbly				For the welcome toast only.
3.2.2. Liquor & Cocktails	incl.(1)			
3.2.3. Soda, water, coffee	incl.(1)			From home bar and pantry.
3.3. Servers and barperson		$ 200		

(1) included

or need to go deeper into a specific cost component of a line item, it will make practical sense to break down only the most cost significant line items into its components. Decision on the level of detail will depend on the magnitude of the specific cost and the size of the project.

The *Project Cost Breakdown* (PCB) is prepared for the purposes of budgeting, cash flow analysis, progress monitoring, and price breakdown for project progress payment claims. Depending on project complexity, several formats can be used. Table 5–3 and Table 5–4 provide examples of preliminary cost breakdowns for the dinner party project compiled with costs obtained from the caterer, servers, barperson, and the entertainer.

PCB and Project Price Breakdown

Unless it is a *unit price contract*, a price breakdown will be needed to make progress payment claims during the execution of a project. A typical example would be *Lump Sum—Turnkey Design and Build Contracts*, where the contractor may not be required to disclose all price details in his/her winning bid. The PCB can easily be turned into a *Project Price Breakdown* by applying the contractor's *profit margin* and any other particular service fees that may be specially requested by the owner to it. Creating a price breakdown based on the structured activities of the PCB provides consistency in scope definitions and in monitoring their scheduled and achieved progress.

Different Perspectives for PCB

Different perspectives of participants may require different PCB approaches for the same product/service. In the previous Dinner Party example, the cost of the main course is not the top priority for the party host. However, the caterer may have a different view. If hundreds of similar main courses are produced on a regular basis, looking into that cost with closer attention would be essential for the caterer. For each participant, the priorities and the related tasks' costs will be different. In this example, the main course may be the top priority for the caterer whereas the quality of the sound system may be the top priority for the entertainer, each task having totally different costs for the caterer and the entertainer. Meanwhile, for the host, the top priority may be selecting the right caterer and the entertainer, a task which may not even incur a cost other than the host's time.

Theoretically any project scope can be broken down to infinite levels. The manager needs to make a judgment call where to stop in establishing what level of detail will be sufficient for his/her planning and control purposes. One way to make that judgment call would be by determining the significance of that cost when compared to other costs (i.e., cost prioritization). Integrating the available cost data with the WBS provides a good indication of what needs to be further detailed and prioritized. However, the tasks that may have minimal cost impact

Hotel Development Project (top level WBS)

1. Development
  1.1 Project Management Cost
  1.2 Owning Company Costs
  1.3 Licenses, Permits
  1.4 Finance Costs

2. Land
  2.1 Purchase / Lease
  2.2 Infrastructure improvements

3. Construction
  3.1 Design
  3.2 Construction
  3.3 FF&E

4. Operations
  4.1 Pre-Opening
  4.2 Working Capital

5. Contingencies

3.2 Construction (WBS Levels 4 & 5)

1 SITE PREPARATION AND INFRASTRUCTURE
  1.1 Site Preparation
  1.2 Infrastructure Systems
  1.3 Site Access
2 SITE DECORATION
  2.1 Decorative Hardscape
  2.2 Pools / Slides
  2.3 Decorative Water Features / Fountains
  2.4 Site Furnishings
3 LANDSCAPE AND IRRIGATION
  3.1 Irrigation System
  3.2 Fine Grading & Soil Preparation
  3.3 Plants & Other Elements
  3.4 Maintenance During Construction
4 BUILDING STRUCTURE
  4.1 Foundation
  4.2 Superstructure
5 BUILDING ENVELOPE/PERIMETER
  5.1 Roof System
  5.2 Prefabricated Exterior Walls
  5.3 Blockwork
  5.4 External Wall Insulation
  5.5 Façade Systems
  5.6 Plaster Rendering
  5.7 Windows
  5.8 Skylights
  5.9 Exterior Paint
  5.10 Louvres and Screens
  5.11 External Doors
  5.12 Other Perimeter/Envelope Finishes
  5.13 Exterior Building Signage
  5.14 Exterior Building Lighting
6 INTERIOR FINISHES
  6.1 Internal Partitions
  6.2 Ceilings
  6.3 Plaster Rendering
  6.4 Floors
  6.5 Walls
  6.6 Ceilings
  6.7 Internal Doors
  6.8 Ironmongery
  6.9 Bathroom/WC Accessories
  6.10 Specialty Systems
  6.11 Interior Signage
  6.12 Interior Landscape
  6.13 Lift Cabs
7 MILLWORK
  7.1 Floors and Platforms
  7.2 Wall, Columns & Partitions
  7.3 Ceilings
  7.4 Counters
  7.5 Cabinets
8 HEATING, VENTILATION & AIR-CONDITIONING
  8.1 Fan Coil Units
  8.2 Ventilation Fans
  8.3 Exhaust Fans
  8.4 Kitchen Exhaust Fans

8 HVAC Cont'd.
  8.5 Fire Dampers
  8.6 Volume Air Dampers
  8.7 Air Handling Units
  8.8 Water Cooled Package Units
  8.9 Refrigeration Plant
  8.10 Chemical Treatment System
  8.11 Boiler Plant
  8.12 Ductwork and Distribution
  8.13 Piping, Valves & Fittings
  8.14 Controls
  8.15 Test and Commission
9 HIGH VOLTAGE SYSTEM
  9.1 Generator-Set
  9.2 Transformer
  9.3 Distribution Boards
  9.4 Metering
  9.5 Sub Distribution Boards
  9.6 Installation and Cabling
  9.7 Lighting
  9.8 Dimming System/Controls
  9.9 Emergency Lights & Signs
10 LOW VOLTAGE SYSTEM
  10.1 Telecom System
  10.2 Audio Visual Systems
  10.3 MATV System
  10.4 Interactive TV System
  10.5 Security System
  10.6 UPS
  10.7 Computer System
  10.8 Electronic Door Locks
  10.9 BAS
11 PLUMBING
  11.1 Cold Water System
  11.2 Hot Water System
  11.3 Waste Water System
  11.4 Pools/Water Features
  11.5 Gas System
  11.6 Drainage
  11.7 Kitchen & Laundry
  11.8 Sanitary Ware & Fixtures
  11.9 Utility Metering
12 FIRE PROTECTION
  12.1 Fire Protection Equipment
  12.2 Fire Alarm Detection
  12.3 Test & Commision
13 LIFTS & ESCALATORS
  13.1 Passenger Lifts
  13.2 Service Lifts
14 MISCELLANEOUS
  14.1 Permits & Fees
  14.2 Survey and Control
  14.3 Test & Inspections
  14.4 Mock-up Room
  14.5 Owner's PM Office

3.2.4.2 Building Superstructure

1. Floor 1 Slab
2. Floor 2 Slab
3. Floor 3 Slab
4. Floor 4 Slab
5. Floor 5 Slab
6. Floor 6 Slab
7. Floor 7 Slab
8. Floor 8 Slab
9. Floor 9 Slab
10. Floor Slab

3.2.4.2.1 Floor 1 Slab
1. Forms
2. Concrete
3. Rebar
See tables 6-2 & 6-5 for details of these tasks.

Figure 5–2 WBS example demonstrating how a floor slab activity of a typical hotel building development project is cascaded down from the top level. Further levels of this WBS is used as the basis for Appendix A.2.

Hotel Development Project WBS (Levels 2&3)

1. Development
1.1 Project Management Costs
1.2 Owning Company Costs
1.3 Licenses, Permits
1.4 Finance Costs

2. Land
2.1 Purchase / Lease
2.2 Infrastructure improvements

3. Construction
3.1 Design
3.2 Builder's Work
3.3 FF&E

4. Operations
4.1 Pre-opening
4.2 Working capital

5. Contingencies

3.3. FF&E WBS (Levels 4 &5)

3.3.1 Furniture & Furnishings
  3.3.1.1 Guest Rooms
  3.3.1.2 Public Areas
  3.3.1.3 Art Program
  3.3.1.4 Interior Signage

3.3.2 Major Equipment
  3.3.2.1 Kitchens and Bars
  3.3.2.2 Laundry

3.3.3 Operating Equipment
  3.3.3.1 Silverware
  3.3.3.2 Chinaware
  3.3.3.3 Glassware
  3.3.3.4 Linen
  3.3.3.5 Uniforms

3.3.4 Special Equipment
  3.3.4.1 Management System
  3.3.4.2 Office Equipment
  3.3.4.3 Material handling Trucks
  3.3.4.4 Cleaning Equipment
  3.3.4.5 Dining Room wagons
  3.3.4.6 Shelving and Lockers
  3.3.4.7 Vehicles
  3.3.4.8 Banquet Equipment
  3.3.4.9 Recreational Equipment
  3.3.4.10 Guest Room Accessories

3.3.5 Auxillary Equipment
  3.3.5.1 Kitchen & Stewart Utensils
  3.3.5.2 Dining Room Accessories
  3.3.5.3 Engineering Tools & Equipment
  3.3.5.4 Housekeeping Utensils
  3.3.5.5 Miscellaneous Equipment

3.3.3.3. Glassware (Level 6; Work Package Level)

1 Tumbler - guest bathroom
2 Tumbler - minibar
3 Highball - minibar
4 Stemmed glass - minibar
5 Champagne glass - minibar
6 Water glass
7 Red wine glass
8 White wine glass
9 Wine glass - vintage
10 Wine glass - multipurpose
11 Champagne flute
12 Champagne coupe
13 Juice glass
14 Highball
15 Old fashioned
16 Liqueur glass
17 Sherry / port glass
18 Grappa glass
19 Brandy snifter
20 Martini glass
21 Cocktail glasses - several types
22 Margarita glass
23 Shot glass
24 Irish coffee glass
25 Glass - smoothies
26 Beer mug
27 Beer glass - Pilsener
28 Beer glass - slever
29 Acrylic stemmed glass
30 Acrylic beer glass
31 Acrylic highball
32 Acrylic smoothies glass
33 Acrylic old fashioned glass
34 Acrylic ice cream / sundae dish
35 Acrylic plate - small
36 Acrylic plate - large
37 Acrylic jug
38 Acrylic salt grinder
39 Acrylic pepper grinder
40 Chinese speciality wine glass
41 Sake cup / glass - cold sake
42 Sake carafe - cold sake
43 Dish - ice cream
44 Dish - speciality dessert
45 Dish - snacks small
46 Dish - snacks large
47 Bowl - decorative small
48 Bowl - decorative large
49 Platter - decorative small
50 Platter - decorative large
51 Plate - decorative small
52 Plate - decorative large
53 Cake stand
54 Carafes/Jugs - several sizes
55 Decanter
56 Glass jars - several sizes
57 Olive oil bottle
58 Vinegar bottle
59 Buffet style juice dispenser
60 Vases - several sizes and shapes
61 Candle holder
62 Ashtrays
63 Accent items

Figure 5–3 Another approach to Hotel Development Project WBS: Example for how the WBS may be detailed by the hotel operator.

but are essential for accomplishing other tasks with significant costs cannot be ignored.

Tables 5–3 and 5–4 depict several tasks that are without assigned costs but that are obviously essential for this example project. Choosing invitees and inviting them may not be the costliest task in this project, but without those guests there will be no dinner party. The same is valid for tasks 1.4., 1.5., and 1.7 of this example project. Similar to *milestone activities* being tasks with zero durations that need to be completed so that some other dependent activities can start and/or finish in scheduling, these tasks can be considered as *Cost Milestones*. These tasks may have negligible costs but other dependent tasks cannot be started/completed unless they are accomplished. Further breaking down those costs into their components and determining the nonmonetary resources allocated for them will be needed to be able to plan and control such cost milestone tasks.

WBS Examples for Different Stakeholders

Hotel property developments, being large projects, involve several stakeholders and key participants. An additional stakeholder, the hotel operator, joins the standard owner-designer- builder trio in these projects, especially when the property is intended to be managed/operated by this stakeholder's brand. Each major stakeholder develops their WBS from their own perspective. The entire WBS for such a project may take several hundred pages. Figure 5–2 and Figure 5–3 provide two samples, one for a typical reinforced concrete floor slab of the building superstructure and one for hotel operating equipment—glassware—are provided to save space. These are samples of a WBS from an owner's and from a hotel operator's perspectives.

Appendix A provides budget formats for different stakeholders on the basis of their consolidated WBSs.

In Summary

1. The first step in project management is planning.
2. Planning starts with identifying what needs to be done to achieve project goals.
3. The WBS identifies and details what needs to be done.
4. The process of creating a WBS and its uses can be summarized as:

    4.1. Breaking down the entire scope creates WBS.
    4.2. The most detailed level of a WBS is work packages (i.e., activities).
    4.3. The activities are:

        4.3.1. Estimated and assigned costs that build up to project cost budget.
        4.3.2. Assigned durations based on cost data which is used for scheduling them.

    4.4. Once both cost and time constraints are satisfied, both cost budget and schedule are used to prepare:

        4.4.1. Project expense schedule: costs of activities spread out to their scheduled timelines.

        4.4.2. Progress payment claims—revenues—schedule: contract priced activities spread out to their scheduled timelines.

        4.4.3. Project cash flow: revenue—expense balance spread out to scheduled timelines, with considerations of leading expenses and revenue lagging due to processing.

        4.4.4. Resource schedule: plotting required resources over the project timeline.

5. Using the same activity definition base provides consistency and easier transferability of information in between the previously mentioned applications.

6 Cost Estimating

Introduction

Cost estimating is the process of identifying the costs of what needs to be expended out to accomplish those goals. It is the process of determining both elements of an activity cost and assigning appropriate resources to complete that activity within its cost and time constraints. This chapter covers types of cost estimating used in the construction industry. A detailed process of preparing a bottom up estimate, including duration calculations for labor and equipment, is provided. The effect of differing priorities of main stakeholders for delivery system and contract pricing combinations is discussed. A detailed example demonstrating each step in preparing a task estimate, testing it against the task's time constraints, and updating of the Work Package Dictionary is presented. Estimating challenges for various types of overheads is explained. Creating a resource-based quantity schedule based on the project work package dictionary is offered as a holistic approach for planning, organizing, procuring, and controlling project resources. Cost estimating is a sophisticated job, requiring judgment calls based on experience. In putting together the cost for a project, estimators have to consider direct and indirect costs of known and known-unknown conditions that complicate those judgment calls.

Cost Estimating in the Construction Industry

Cost estimation is a significant effort in the construction industry. Owners as well as general and sub-contractors prepare cost estimates for their own purposes. Bill of quantities (BOQ) and cost estimates prepared by the Architect or Engineer (A/E) serves as an owner's cost reference. When preparing to receive bids from contractors, a project scope definition is supplemented with all available documents such as drawings, specifications, benchmarks, and various other sources and references which are collectively called the *tender documents*. By making such documents—most commonly with the exception of the A/E estimated prices—available for bidders, the owner secures that all bidders prepare their bids on the same scope and quantity basis. Providing the BOQ to the bidder prevents errors and omissions of some parts of the scope by mistake. However, during the project

DOI: 10.1201/9781003172710-6

execution, some benefits of providing a BOQ to potential contractors may turn out to be counterproductive if site conditions, quantities, and types of indicated work differ from what has been foreseen at the time of contract award. Such differing conditions cause significant delays and cost overruns for contracts entered through such practice.

In a bid where a contractor prices out a provided BOQ and is granted the contract, that BOQ becomes a part of *Contract Documents*. The contractor—and its appointed PM—is in a position to verify the owner-provided BOQ for type of work specified and for quantities against actual quantities carried out on site. It is very likely that the contractor's estimator will not verify the scope completeness and provided quantities while preparing for the bid. The estimator's focus is on the unit prices at that phase. If/when the contract is awarded, the contractor's PM will then be in a position to verify that not only the unit prices are good for the work described by the contract documents but also the amount of work given in the BOQ matches the actual quantity of work to be carried out. If a significant discrepancy is found, higher management needs to be informed to implement the contractually prescribed change order process. What happens after that depends on several factors and stakeholders' company culture and is beyond the scope of this book.

If the type of work and the quantities provided by the contract documents are in line with the PM's findings, the project team may use the provided BOQ, which may be converted to the project WBS as the first step in their planning process and continue with selecting the construction method, assigning/allocating resources, and proceeding with both cost estimating and scheduling.

If a BOQ is not provided (i.e., in a lump sum priced bid), the contractor's estimator has to prepare the BOQ (i.e., the WBS and a quantity take-off (QTO)), assign durations based on method and resource allocation assumptions, and calculate first the cost and then—with the input of general margin and profit—find the bid price for the project. Again if the contract is awarded, the contractor's appointed PM will have to verify all information provided and complete the process as shown in Figure 6–1.

Types of Estimates and Estimate Accuracies

Several types of estimates can be used depending on the level of detail available on the scope and methodology of the project. As expected, availability and detail of information increases as the project moves through its phases. For the manager and/or estimator, very limited project information will be available at the conceptual phase and the full information should be available at the closing phase. Although no project will be exactly the same, *lessons learned* from other similar projects are the best starting references for any person who is in a position to make an estimate on a new project. The limitations on the available information and the similarities/differences of projects are the reasons for a wide range of inaccuracies for estimates.

At the *conceptual phase* of the project the "concept" is pretty much all the information available for the project. This is the time when at least a preliminary

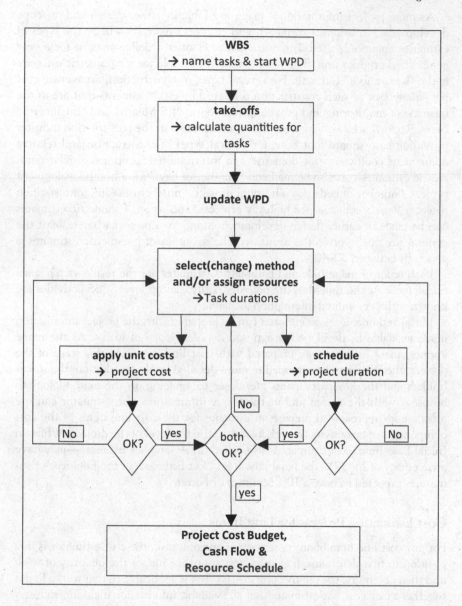

Figure 6–1 Estimating process and cost/schedule iterations.

feasibility needs to be conducted to make the "go ahead" decision. An estimate needs to be in place to carry out that preliminary feasibility. That estimate most probably will be an *Order of Magnitude* estimate. Also referred to as a *Schematic Estimate*, the order of magnitude estimates are prepared with 50–100% accuracies. Since this is a very low accuracy, further estimates are essential to refine that accuracy as more project information becomes available.

As more project information is made available by either design and/or scope development, *parametric*, *assemblies*, and *engineer's estimates* will be the types of estimates commonly used. *Parametric estimates* offer a dollar amount for a very generalized product unit. The most common example for a parametric estimate is the "square foot" estimate. For several types of construction, an average cost per square foot of such construction is offered by establishments that are in the business of monitoring and providing such data. "RS Means" and "Engineering News Record" are two very well-known institutions in the construction industry providing base square foot costs for several types of construction and relative adjustment coefficients for domestic and international locations. Similar parametric estimates can also be made on "number of bays" for a hotel development project, "number of beds" for a hospital project, "number of seats" for a stadium project, "lane x mileage" for highway projects, and so on. Parametric estimates can be made in earlier design development stages of a project when at least the general size of the project is available. The accuracy of parametric estimates is generally between 30–40%.

Both order of magnitude and parametric estimates are the results of *top down calculations*. At the time they are prepared, only the top level WBS activities are known with very limited information available.

Initial and final *line item estimates* can be prepared with the project information made available by developed design and detailed scope of works. As the name implies, these estimates are prepared with cost line items. At this stage of the project, the WBS can be extended to more detailed levels. In addition, the stakeholders and the key participants are closer to agreeing on the methodology to be used. With the design and methodology information, the estimator can use a *bottom up* approach to prepare an estimate for the cost line items of the cost components of an activity to work its way to the top level of the project. While an Initial Line Item Cost Estimate—also called an Assemblies Estimate—may have an accuracy of 25–30%, the Final Line Item Cost Estimate, or the Engineer's Estimate, is expected to have a 10% accuracy or better.

Cost Estimating Process for Line Items

For any cost line item of any cost component of a task, the cost estimator is in a position to first determine (i.e., measure, count, or judge) the quantity of work and then determine the relative unit cost(s) that is applicable to that work. To put together a task cost, the estimator uses all available information including industry references, company records from previous similar projects, his/her own experience, and subject matter experts within or outside the organization.

The estimating process is finding the answers to following questions:

a)   *what* (i.e., scope line items),
b)   *for what quantity,* and
c)   *for how much* to get the job done.

In this statement *what* relates to any task cost component line item, *for what quantity* relates to its required quantity, and *for how much* relates to the applicable unit cost for that specific line item quantity.

As the WBS tasks at certain detail levels eventually add up to constitute the entire scope of work for a project, the costs assigned to sub-tasks (i.e., cost component line items) add up to that task's cost and the summation of task costs at all levels add up to project's estimated cost.

The process starts with preparing *quantity take-offs* (QTO) for tasks that define the entire project scope. If there is already a WBS available (i.e., BOQ for a DBB project), the estimator can start determining quantities for WBS tasks; otherwise, the estimator will have to start the estimate with preparing that WBS.

Quantity Take-Offs (QTO)

Preparing QTO is counting, measuring, and calculating the quantities of cost line items for all relevant cost components. Some of these quantities can be obtained digitally where project documents are digitized. Specifically developed software to count and/or measure quantities from document hard copies is also available for non-digitized, hardcopy drawings. The last resort to count and measure is doing it manually and most probably by using several colors of markers to highlight different task-related items on drawings. QTOs are prepared for materials, labor, and M/E.

QTO for Materials

When drawings and other *design documents* are available, the QTO for materials are obtained straight from those documents. Material QTO is simply counting, measuring, and calculating the specific line item materials. If that line item is one type of lighting post, the total number can be obtained by counting them from exterior lighting drawings. Some other quantity may require obtaining dimensions from drawings and calculating areas, volumes, or weights. If the line item is painting interior walls, the area that needs to be painted needs to be calculated by measuring the linear footage and the height of walls along with void dimensions such as for doors from relevant drawings. Once the total area is calculated, it can be used to calculate the paint quantity as well as the quantities (i.e., durations) for labor and M/E.

Labor and M/E QTOs and Durations

The quantities for labor and M/E are durations of these resources required to complete a certain amount of work for the subject task. Once the quantities of work are determined, these durations can be calculated by using *performance data* that relate to these resources. The performance of labor and M/E resources are expressed in two types of *output rates*. One type of output rate provides the quantity of work a resource can produce per unit time (i.e., 12 sq. ft. per hour). Alternatively, the amount of time it takes to produce a unit of work can be used as well (i.e., 5 minutes per sq. ft.). Both performance data need to be based on a specific selected method that will be used to carry out the task. Table 6–1 provides such output data for the labor cost component for a typical wall painting task. Once the method is determined for this task, the M/E quantity can be calculated by using the same output data. As an example, if the *spraying* method is used, the spraying equipment usage will be calculated by using the labor hours needed to spray the required quantity of wall area.

The M/E manufacturers as well as references such as RS Means provide performance rates. Such references as well as data from other similar projects and subject matter expertise should also be used as may be needed.

Establishing the task component quantities (i.e., material quantities, labor, and M/E durations), the cost estimator applies the appropriate unit costs and calculates the cost for a task and sums up all task costs toward calculating the project cost. While task-related indirect costs can be calculated with this data, to determine the indirect project costs (i.e., project O/H), the estimator may need to know the overall project duration. The project duration cannot be determined by summing up the task durations. Project duration is the longest path of task durations which are sequenced according to the methods used. This scheduling process is the job of the scheduler and not the estimator. For an initial estimate for the O/H costs, the estimator will need to assume a project duration which then must be verified when the project schedule is established.

The duration of a task depends on the method used and the resources allocated to accomplish that task. Once the quantities, labor, and M/E performance data are known, the durations of the tasks can be determined. The durations of tasks is a helpful input for the scheduler, who may/may not agree with the estimator's assumptions in determining them. It is the scheduler's job to verify such durations and sequence them as per method(s) used to determine the project duration. Eventually both cost and time requirements are to be met. Figure 6–1 demonstrates the schematics of this process where such iteration continues until both cost and time constraint are fulfilled.

Estimate Preparation Duration and Accuracy

Another factor affecting the cost estimate accuracy is the amount of time allowed for the estimator to spend on preparing the estimate. On several occasions project stakeholders may need project cost estimates "quickly". It is not unusual that the final cost of a huge project will be requested within minutes where the only information provided is just a *concept*. For even seasoned professionals, providing such information on such limited information and so quickly creates considerable concern on their own professional credibility and liability. Cost estimates can be prepared on the run or even on a paper napkin over a beverage. Limits on preparation time may negatively impact accuracies by multiple times. When tasked with preparing a cost estimate, establishing the amount of time allowed for that study

Table 6–1 Interior drywall paint task material quantity and labor data

Cost component & line item description		Quantities for Methods		
Cost component	Line item	Brush	Roller	Spray
MATERIAL				
COVERAGE PER GALLON OF PAINT	Priming	450 ft^2	425 ft^2	550 ft^2
	2 finishing coats	500 ft^2	475 ft^2	550 ft^2
LABOR				
OUTPUT PER HOUR	Priming	200 ft^2	300 ft^2	500 ft^2
	2 finishing coats	200 ft^2	300 ft^2	500 ft^2
HOURS NEEDED TO PAINT 100 ft^2	Priming	0.5 hrs	0.3 hrs	0.2 hrs
	2 finishing coats	0.5 hrs	0.3 hrs	0.2 hrs

and calculation is highly advisable. In return, it is a good practice to include the expected accuracy level with any cost estimate.

Expert Opinion in Estimating Costs

As projects get more complex with sophisticated and unique deliverables, costs for such deliverables cannot be estimated by simply using existing data. To evaluate and adapt limited cost data to be applied to the project tasks at hand, the estimators may need to consult with experts. Although the outcome of the expert opinion may easily be a best guess, that could be better than nothing if nothing else is available. Such cases need to be handled with utmost attention so that neither the owner nor the contractor is unfairly held accountable for the results later on during the project.

Using a *provisional sum* is one practice that can be used for such cases. A provisional sum is a best cost estimate and/or a not to exceed amount prepared with what is known and/or guessed at the time of entering a contract. That estimate is expected to be revised and developed as parties know and learn more as the project proceeds.

Step-by-Step Cost Estimating Process

The estimating process schematics in Figure 6–1 outlines how an estimate is developed and its cost and duration results verified with project targets before producing project budget, schedule, cash flow, and resource schedules.

The first step in estimating is preparing the WBS unless it is provided. If the provided WBS tasks are in a form the estimator can use to develop an estimate, then a further WBS may not be necessary. Either way the estimator needs a WBS to start the process.

The second step in estimating is determining the quantities for all cost components. The quantities for material, labor, and M/E—and some O/H costs—are calculated by measuring and calculating them. Measuring and calculating quantities for line items of tasks is defined as preparing quantity take-offs (QTO). A generally accepted checklist for preparing QTOs includes the following:

Material take-off checklist:

- Measure and count where practically possible. Add consumables and other miscellaneous items (i.e., those that are not measured/counted) as per manufacturer recommendations, professional guides/manuals, data from similar projects, and estimator's experience.
- Allow for wastage considering the method used. Also consider storage, staging, horizontal/vertical transportation, and handling at the work site.
- Allow for **attic stock** (i.e., saving spares for future use considering the same material may not be available) as well as for damages that may occur well after completion of the work.
- Consider provisions for protection and maintenance of completed work until other related tasks in close proximity are completed as well.

Labor and M/E take-off checklist:

- Using **output rates** for the selected method to be used (or labor-hours and/or M/E-hours required to produce a certain amount of work) determines total labor hours and M/E hours required to complete the task.
- If the used output rates do not already include these, allow/adjust for down time, productivity, climate-weather conditions, workplace conditions (lighting, air quality, accessibility, etc.), and any other condition that may adversely impact the assumed output.

The third step in preparing an estimate is to check task durations and make the decision to keep or change the selected method for carrying out the task being estimated. If the results are not favored, then the estimator may choose to try another method. The process for checking the task durations is outlined here:

- Select a number of workers to be assigned to this task (a crew) and using the assumed work hours per day, workdays per week, and other holidays and non-working times, calculate the task duration. For M/E, while the M/E hours can be calculated directly by using the task quantity and M/E output rate, the operator (i.e., direct labor cost) and support and maintenance (i.e., indirect labor cost) labor costs also need to be considered and included.
- If in doubt, double check the duration of the same task by implementing a different method but make sure its cost is not significantly different.

The fourth step in estimating is two simultaneous steps. These two steps are considering the task cost and duration at the same time so that both project cost and time constraints are satisfied:

- Evaluate the resulting duration: consider adjusting crew size, number of crews (number of M/E to be used), overtime and/or holiday work to adjust as needed.

For cost analysis:

- Apply unit costs for materials and hourly rates for labor and M/E including surcharges such as O/T and determine the task cost.
- Tentatively evaluate if the task cost is acceptable. This step may need to be revisited after all task costs summing up to project cost are calculated.

For duration analysis:

- Tentatively evaluate resulting duration. How this duration will impact the project schedule may not be easy to predict without the project schedule being completed. This step may need to be revisited after scheduling is completed and provides the project duration.
- If either the cost or duration initial evaluation is unsatisfactory or doubtful, revisit the selected construction method. Search for an acceptable alternative method that may produce better cost and duration results.

Finally, if and when both cost and duration results are satisfactory, they may be included in the project cost budget and used for cash flow and resource schedule preparations.

Estimating is an iterative process. First it has to deal with subparts of a scope and second, it targets to optimize the cost and the duration while the entire picture is not available and the whole process is not complete. It may take several passes and revisits to assumptions and selections before an all acceptable solution—if that is ever possible—can be reached.

Project Delivery Systems, Contract Pricing, and Cost Estimating

Several delivery methods and contract pricing options may lead to differing approaches of main stakeholders to project WBS, cost, and duration. The list of main stakeholders may be extended to include the owner's technical and/or contractual management representation, specialty contractors, operator's design and technical representation, and suppliers/manufacturers of specialty equipment and products.

While the process of cost estimating remains the same, stakeholders' targeted precision level and priority ranking between the estimated project cost and duration varies depending on the chosen delivery method and the contract pricing. In addition to the primary triple constraints, qualitative constraints such as quality of workmanship, professionalism, and customer satisfaction may also be in the picture as priorities of stakeholders for some delivery method/contract type

Table 6–2 Priorities of stakeholders for select delivery method—contract pricing combinations

Delivery method	Unit price contract priorities of		Lump sum contract priorities of		Cost plus contract priorities of	
	owners	G/Cs	owners	G/Cs	owners	G/Cs
DBB	lowest cost	best unit price	lowest cost	1. scope 2. cost	n/a	n/a
CM	lowest cost	best unit price	lowest cost	1. scope 2. cost	1. quality 2. time	client satisfaction
DB	n/a	n/a	scope quality time	1. time 2. scope 3. quality	1. quality 2. time	client satisfaction
DBOT	n/a	n/a	time	1. time 2. cost	n/a	n/a

combinations. Table 6–2 depicts the priority rankings of owners and contractors for most common delivery method and contract pricing combinations.

Although cost and time planning processes remain the same, the indicated priorities drive the estimators and schedulers to pay closer attention to them. In a unit price contract to be awarded by using DBB, the estimator needs to put his full focus on the unit cost he/she will use to price out the owner-provided cost line items. In this case the quantities are provided and are expected to change as the work progresses. Time is a concern from a contractual delay–penalties perspective but the main goal is to bid the lowest price. On the other far end, for a Cost Plus (i.e., an incentive) contract and DB delivery combination, the owner depends on the know-how and skills of the contractor to design and build a complete and unique product at an acceptable price level and within a certain timeline. This sets the focus of the estimators on the scope coverage and quality rather than quantity and competitive pricing. Table 6–2 tabulates the other option combination priorities for owners, construction management companies, and contractors.

Technology and Reference Data for Estimating

Estimating technology has advanced significantly during the past two decades. Several software options are available for WBS, take-off, and cost/bid preparation. In addition, Building Information Modeling (BIM) has increasingly been the choice of architects and engineers, allowing them to incorporate various information such as product types, quantities, and cost data into their digital designs. Existing cost database companies expanded their geographic coverage areas and have included more construction types in their databanks.

However, even with current substantial improvements, a generic software output on its own is very unlikely to be sufficient for any project stakeholder to commit to a contract based on that output. Since every project is by definition unique, some adjustments in light of the project specifics should be expected. These adjustments may come in the form of labor efficiency, site conditions and location, client requirements, applicable regulations and code, and many similar others. Such adjustments need to be applied with some judgment call by the estimators and/or PMs. One additional form of such adjustments comes under the general term *contingency*, which is analyzed in detail in Chapter 7.

Updating WPD With Estimated Cost Data

When cost estimating is complete with duration/schedule verification, the WPD needs to be updated with the latest and additional information obtained. By updating the WPD with the estimated cost data (i.e., quantities, units of quantities, unit costs, and total costs for all component line items), the PM will be able to obtain the totals—both quantities and costs—for all materials, labor trades, and M/E for all tasks. Since this set of information is the result of an estimation at the start of a project, it will serve as the reference baseline for actual results, both at the work package level and for the overall project during its execution and closing phases. Consolidating all relevant information in the WPD also allows elimination of hidden inconsistencies between information generated by different sources.

When the WPD of the project tasks are updated with estimated quantities and costs, this information can be sorted on the basis of materials, labor trades, machinery-equipment requirements, and other direct and indirect costs. The project totals of specific resources of tasks can be obtained by sorting and summing them up on a specific resource basis. What could be named as a **Project Resource BOQ (RBOQ)**, this compilation of resource cost data (i.e., quantities, unit costs, output rates) is a very useful tool for planning, organizing, procuring, and controlling these resources. Figure 6–2 demonstrates the schematic for this conversion process.

WPD and Project Resource Schedule (RBOQ)

When completed, the project WPD becomes the main reference for resources, costs, and durations for project tasks. It is worth re-mentioning that along with the unit and total costs, the resource quantities used in estimating and scheduling are extremely valuable information for managing a project. This information is not only needed for planning but also sets the reference for controlling the project cost as well as the schedule. With the recompilation of the WPD, it can be turned into a *resource bill of quantities* (RBOQ) which can be used similarly to a BOQ. Figure 6–2 presents the schematics of how a project resource schedule can be obtained by re-sorting an activity-based WPD on the basis of resources grouped under cost components.

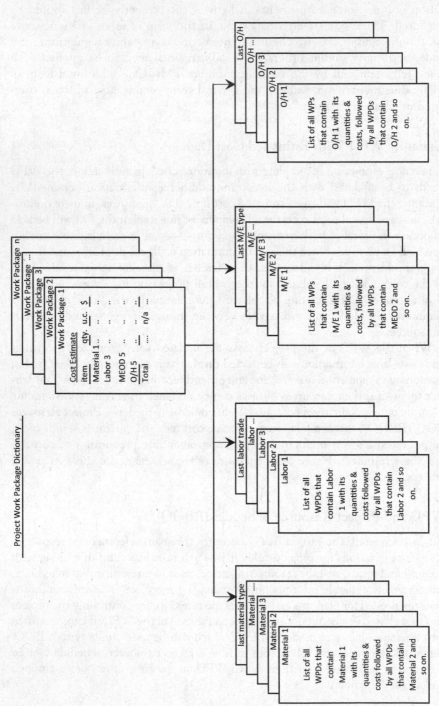

Figure 6–2 Task component line items can be re-sorted to calculate the critical resource totals for the entire project. Figure schematicizes re-sorting "n" number of tasks for varying number of their cost component resources. Selecting only critical resources will shorten the list to focus on.

Example: Hotel Building Superstructure

Quantity Take-Offs, BOQ, WPD Update, and Project Resource BOQ

The reinforced slab for floor 1 of the superstructure of a hotel development project (Figure 5–2), will be used to demonstrate the process depicted in Figure 6–1. It should be noted that in this example one part of a building project, the "superstructure" of that building is considered as a project of its own. Making that consideration simplifies exemplifying the process. In addition, each floor slab of the building superstructure is assumed to be identical, again for simplifying the example.

The subject building's superstructure comprises 10 identical floors (i.e., 10 slabs) with identical reinforced concrete design. Each slab is 210' long, 50' wide, and has:

- two 14' by 18' emergency stairwells,
- two 8' by 8' elevator shafts,
- eleven 2' by 3' service shafts,
- solid slab thickness of 8",
- forty-eight 8" by 16" columns,
- 8" thick shear wall around elevator shafts, and
- the floor clearance is 12'.

STEP 1. Prepare (or Check if Available) Quantity Take-Offs

Preparing quantity take-offs either manually or using any available technology will be the first step in cost estimating this work package. Since technology advances as we breathe, there are several options for preparing a take-off. However, regardless of how it is done and who does it, preparing a take-off is measuring the quantities of the cost components of the tasks in units that are in line with what is accepted in the industry. In the construction industry, formwork is measured in units of area, reinforcement steel in units of weight, and concrete in units of volume. Table 6–3 shows what is a typical quantity take-off for such building floor concrete work based on the information provided.

STEP 2. Update the WPD for This Activity

At the end of the estimating process, the WPD for this estimated activity will be updated as shown in Appendix B.

STEP 3. Assign Resources and Determine Durations

Applying how many labor hours it will take to get the job done is one of the most difficult assessments in the entire estimating and scheduling process. There is quite a bit to consider here. The method used, the selected equipment and its condition/efficiency, labor skill level, labor efficiency under the site/weather

Table 6–3 Sample quantity take-off for hotel building superstructure floor slab

3.2.4.2. SUPERSTRUCTURE (refer to WBS of Figure 5–2)

WBS No	Description	length	width	height	times	quantity	unit	remarks
3.2.4.2.1.	**SLAB for FLOOR 1**							
3.2.4.2.1.1	**FORMS**							
	Columns	8"	16"	12'	48	2,048.00	sq.ft.	
	Elevator walls	48'		12'	2	1,052.00	sq.ft.	
	Slab deck gross incl. work area extension	222'	62'		1	13,764.00	sq.ft.	extend 6' at each side
	Less elevator shaft	8'	8'		2	(128.00)	sq.ft.	deduct for
	Less service shafts	2'	3'		11	(66.00)	sq.ft.	concrete
	Less stairwells	18'	14'		2	(504.00)	sq.ft.	quantity
	Safety railing	222'	62'		2	568.00	ft	
	Slab exterior sides	210'	50'		2	540.00	ft	
	Elevator shaft front	8'			2	16.00	ft	
	Service shafts	2x2'	2x3'		11	110.00	ft	
	Stairwells	18'	14'		2	128.00	ft	
3.2.4.2.1.2	**CONCRETE**							
	Columns	8"	16"	12'	48	19.00	cu.yd.	
	Elevator walls	48'	8"	12'	1	14.00	cu.yd.	
	Concrete for net slab area	9882 sq.ft.		8"	1	244.00	cu.yd.	
3.2.4.2.1.3	**REBAR**							
	Columns	19 cu.yd.@0.075T/cu.yd.				1.43	tonnes	
	Elevator walls	14 cu.yd.@0.065T/cu.yd.				0.91	tonnes	
	Slab	244 cu.yd.@0.065T/cu.yd.				15.86	tonnes	

conditions, previous experience on a similar work, and group–team/management efficiency are some of the factors that account for the overall result.

If available, adapting some historic data from a previous experience/project would be a good starting reference. Otherwise, after careful review and consultation with the team and with those who might be more knowledgeable, a judgment call will be made. Table 6–4 tabulates those labor-hours that would apply to the units indicated. For sake of this example, further breakdown of the *trades* of labor is not provided in Table 6–4.

Obviously, the information provided in Table 6–4 is not practicably usable. Rather than having one laborer spend 2,297 hours (i.e., 287 labor-days) to install the formwork for one slab will not serve the purpose. A crew of formwork carpenters, scaffolding erectors, and helpers (i.e., semi- and unskilled workers) will be formed to complete this job at a reasonable cost and duration. The same applies to rebar and concrete tasks.

To determine the duration of the activity, the total labor—and other—resource hours will first be converted to days. Then a crew size including skilled, semi-skilled, and unskilled workers will be established. Generally, the number of workers

Table 6–4 Labor quantity (hours) take-off for WBS task 3.2.4.2.1 Slab

Task	Quantity	Unit	Labor-hours/unit	Total hours
FORMS				**2,297**
Columns	2,048	ft^2	0.17	348
Walls	1,052	ft^2	0.11	116
Deck	13,764	ft^2	0.12	1,652
Edges and railing	1,362	ft	0.045	61
Strip and prepare	L.S.[1]	slab	120	120
REBAR	18.20	tons		**301**
Columns	1.43	tons	20	29
Walls	0.91	tons	20	18
Slab	15.86	tons	16	254
Anchors	L.S.	slab	ls	8
MEP				**56**
Electrical	L.S.	slab	24	24
Plumbing	L.S.	slab	32	32
CONCRETE	277	yd^3		**273**
Columns	19	yd^3	1.46	28
Walls	14	yd^3	0.95	13
Slab	244	yd^3	0.85	207
Finishing	13,000	ft^2	0.001	13
Curing	L.S.	slab	12	12

[1] L.S.: Lump sum

in one crew is limited to 10 ~ 20 people depending on the experience of the team foreman. The team foreman works under the general foreman who oversees several other team foremen. In this example, the team foremen hours are assumed to be included but the general foreman's hours are not.

Table 6–5 shows the crew size, number of crews, and how many days it will take this task force to complete the task.

There are a number of things that need to be addressed to the assignments and conclusions given in Table 6–5:

- A fraction of a day indicates that either changing the crew size or overtime is required.
- When transferred to the workday calendar, the labor cost may increase for work done during weekend days and holidays.
- Local and/or company rules may not allow daily adjustment of workers in crews.
- Workers' ability to work in more than one trade should be considered in establishing crew sizes and numbers (i.e., form worker being able to tie rebar and/or place concrete).

STEP 4. *Calculate the Cost by Applying Unit Costs to the Quantities*

A good practice at this step is to note the assumptions made and inclusions and exclusions that apply. Table 6–6 shows the details. In addition, obtained cost information such as quotations from suppliers/vendors should be attached/referred to with specifics.

STEP 5. *Check Cost and Schedule Acceptability*

At this point both the cost and the time should be within the targeted limits. In this case, the effort is to determine the cost of this task which should allow the completion of the entire building structure, the sub-structure and the superstructure,

Table 6–5 Converting total labor-hours to duration in days

Task	Total hours	Total labor-days	Crew size (worker)	No of crews	Total crew hours/day	Duration for crew(s) in days
FORMS	2,297	287.1	15	2	240	9.57
REBAR	301	37.6	15	2	240	1.25
MEP	56	7.0	7	1	56	1.00
Electrician	24	3.0	3	1	24	1.00
Plumber	32	4.0	4	1	32	1.00
CONCRETE	273	34.1	15	2	240	1.14

Note: Other than MEP, indicated hours are skilled/unskilled mixed hours that include crew foremen.

Table 6–6 BOQ for WBS task 3.2.4.2.1 Floor slab

Component	Description	Quantity	Unit	Unit cost ($)	Total cost ($)
MATERIALS	**Ready mix concrete** incl. transprt. and pump.	273	yd³	175	**47,775**
	Forms				**8,070**
	Metal props for deck (100 uses expected, 1/100 cost charged)	700	each	2	1400
	4×4"×8' beams (10 uses)	700	each	2	1400
	3/4" plywood (5 uses)	480	each	8	3840
	2×4"×8' wood (10 uses)	300	each	1	360
	Wood planks 2×8"×10' (10 uses)	100	pieces	1	50
	Pegs 1×2×12"	1	box	20	20
	Form oil	100	gallons	6	600
	Nails, screws, tie rods, braces, and other consumables.	1	l.s.	400	400
	Reinforcement				**57,225**
	Rebar delivered cut, bent, and tagged.	18	tons	3,000	54,600
	Rebar wire	500	lbs.	1	625
	Spacers	400	boxes	5	2,000
	Miscellaneous				**4,600**
	MEP	1	l.s.	2,500	2,500
	Anchors	1	l.s.	1,500	1,500
	Plastic moisture barrier	14,000	sq.ft.	0.100	1,400
	Burlap for water curing	14,000	sq.ft.	0.200	2,800
	Water for curing	100,000	gallons	0.004	400
	Total Materials				**117,670**
LABOR	General Foreman[1]	110	hours	60	6,600
	Carpenter[2]	766	hours	40	30,640
	Iron worker[2]	100	hours	40	4,000
	Laborer[2]	1,732	hours	25	43,300
	Electrician	24	hours	45	1,080
	Plumber	32	hours	45	1,440
	Labor Total				**87,060**
MEOD	The site installations, storage/ staging areas, vertical and horizontal transportation equipment, and similar other resources are used but their cost is not included. The general site operation cost will be calculated and added as a general overhead. Only specific hand tools, lighting, power extensions, and other similar tools/equipment provided are included here.	1	l.s.	1,000	1,000
	MEOD Total				**1,000**

(Continued)

Table 6.6 (Continued)

Component	Description	Quantity	Unit	Unit cost ($)	Total cost ($)
O/H	Only task- specific items such as specific gear, cleaning, and repair supplies are included here.	1	l.s.	750	750
	Total Overheads				750
		TOTAL			$ 206,480

[1]: 1 general foreman hours assumed for ~8 hours of skilled trades
[2]: 2 Laborer hours to 1 skilled hours ratio assumed.

within 120 days from its start. There is no target cost, but with the parameters established we would have a good enough sense of what to expect.

In the 3.2.4.2.1 Slab WPD Progress Measuring Metrics it has been established that this task represented 0.02675% of the entire project. Applying that assumption to our estimated cost of $ 206,480 for this task we could extend our estimate to:

$$\$ 206,480 / 0.002675 = \$ 77,143,925$$

as our direct total cost for this hotel development project. Assuming site and head office overhead and profit to be in the order of 40% of the cost, this project may end up having a sale (or bid) price of $108,000,000. Naturally, this percentage will vary on a number of preferences of the company's top management. While the accuracy of this number is arguable, it should be a good indicator of what to expect, especially in comparison with earlier estimates. For this exercise, it is assumed to be acceptable.

Since the concurrence on the time constraint is the other requirement, a simple schedule check will be done to demonstrate the process. The given completion duration requirement for the 3.2.4 Building Structure is 120 days. The building structure includes the Foundation and the Superstructure subtasks. There are 10 floors with columns, walls, and slabs that constitute the Superstructure. Using the durations calculated in estimating the cost for one slab and assuming the foundation will take 15 days to complete, we could prepare the following simple Gantt chart as shown in Figure 6–3. In this chart, the calculated 9.57 days duration for formwork is rounded up to 10 days, 1.25 days duration for rebar is rounded up to 2 days and concrete placement and finish duration is rounded down from 1.14 to 1 day.

According to this simple chart (i.e., Figure 6–3), it will take 12 days to complete one slab and 120 days to complete 10 of them. This is not acceptable since

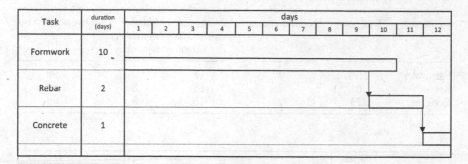

Task	duration (days)	days											
		1	2	3	4	5	6	7	8	9	10	11	12
Formwork	10												
Rebar	2												
Concrete	1												

Figure 6–3 Slab schedule schematics. To save time, the rebar starts 1 day before formwork is completed.

it exceeds the 120-day completion requirement for the entire superstructure, including the foundation.

To comply with that requirement, the superstructure 10 slabs must be completed within 105 days or less. One way of crashing one slab completion duration to 10.5 days or less is to increase the crew sizes. Another way would be by working overtime hours. Two options will be entertained to check the best way to accomplish that duration target of 10.5 days.

It should be noted that the cost estimator may/may not be in a position to conduct the scheduling. In this example the simple scheduling is provided for demonstrative purposes. For more complicated cases the cost planner may need to work together with the scheduler to verify resulting project durations when activity durations are modified.

Option 1. Increasing Crew Size

Initially 2 crews of form, iron, and concrete workers each with 15 workers were used to calculate the number of days to complete the job. Those crew sizes should be revisited in order to complete this job in 10 days. Table 6–7 shows expedited durations of tasks for increased crew sizes. Using these new rounded off task durations, Figure 6–4 verifies that one superstructure slab can be completed in 10 days.

The task durations, rounded to the closest integer days, create some discrepancies with the estimated total hours for these tasks. While rounding *up* will end as time lost, rounding *down* will have to be compensated, either by overtime hours and rates or by increasing crew size. It is generally accepted that at least 4 regular hours will be paid to a worker once they are called in for work. Also, sending off some crew members while the rest of the crew is still working is not a preferred practice. With this in mind, the lost and to be compensated hours are evaluated as follows.

Table 6–7 Revised crew sizes and durations for one superstructure slab

Description	Total hours	Number of crews	Number of workers in each crew	Number of working hours/day	Duration in days
Formwork	2297	2	18	8	7.98
Rebar	301	2	18	8	1.05
Concrete	273	2	18	8	0.95

Task	duration (days)	days										
		1	2	3	4	5	6	7	8	9	10	11
Formwork	8											
Rebar	1											
Concrete	1											

Figure 6–4 Slab schedule schematics with shortened durations. This time rebar starts right after formwork is completed; note a bit of a time contingency there when compared to the schedule of Figure 6–3.

For formwork: 2 crews x 18 workers each x 8 hours/day x 8 days = 2,304 hours is 7 hours more than the estimated 2,297 hours. Similarly, the concrete placement hours for expanded crew size is also 15 hours more than the estimated 273 hours for this work. Considering the size of the operation and several factors that govern the accuracy of this result, these extra 22 hours should not be significant in this context; that would be a *contingency* that will be discussed in the next chapter. For the rebar activity, the same calculation indicates a deficiency of 13 hours to make up to the estimated 301 hours. It can be argued that the subject 13 hours could either be compensated by the available 22 hours from the other activities, or by paying 13 workers one hour overtime. The overtime compensation in addition to the regular pay would generally be 0.5 (overtime premium) x 13 hours x $ 47 (1/3 of $60 skilled rate + 2/3 of $40 unskilled rate combination) /hour = $ 306. Again, in the context of the general effort, this amount can be considered either negligible or as a rounding off error.

Option 2. Keeping the Initial Crew Size and Paying Overtime as Needed

This is a valid option where increasing the crew size is not practically possible, due to shortage of labor, physical conditions on job area, and/or some administrative or legal issue. For the targeted 8-day duration, the rebar workers are to be compensated for: 301 total hours minus 240 hours (30 workers x 8 regular hours/day) = 61 overtime hours. Likewise, the concrete placement crew will need to work 33 overtime hours to complete their work in one day. Using a combined average hourly rate of $47 and an overtime multiplier of 1.5 (i.e., 50% premium per hour), the additional cost of overtime work would be $ 2,209 for the entire slab to be completed in 10 days.

Decision

For this example, option 1 (i.e., increasing the crew sizes) is judged to be sufficient with no need to revise the estimated cost at this point. However, this is a good point to make a remark for contingency considerations that will take place later in the overall cost planning process.

STEP 6. Finalize by Updating WPD

Once the activity costs and durations are verified to be within the project targets, the WPD will be updated. The updated WPD then becomes the reference book for preparing the project cost budget and expense and resource flows.

Remarks on Cost and Duration Correlation for the Example

Whenever a task cost and/or duration is modified, its effect on the other constraint needs to be checked and verified. Although the given example does not demonstrate significant results, such may not be the case when other project tasks are tweaked for duration and cost. It should also be noted that modifying critical activity durations may shift the critical path. For a major project, the impact of delays and/or accelerations can be crucial for making construction method and equipment selection and resource assignment decisions. The estimators and schedulers need to work hand in hand to jointly make those decisions that will be passed on to the PM who will implement them on the job site. In addition, the example used only focuses on direct costs. The indirect costs also need to be checked and verified when the overall project duration is modified, especially for projects that are expected to last years.

Estimating Overhead(s)

The cost of everything that is required to get the job done but which actually does not directly relate to the final product's cost is collectively called overhead (O/H). O/H is an indirect cost that can either be fixed or variable. There are three categories of O/H:

1. Task O/H
2. Project (i.e., site) O/H
3. Head Office O/H

Task O/H

This cost category is task specific and is an indirect cost of the activity. The safety railing around the concrete slab of a high-rise building is an example of such overhead. Such a railing is needed and required by code but will not be a part of the slab when that work is completed. Similar examples would be costs of quality assurance, superintendent, floodlights, rain/wind protection equipment, and safety signs/barriers.

Project O/H

This category of O/H covers the costs of all site installations and services that are required for performing the project activities but are not a part of any specific activity. The safety fence around the project site cannot be directly related to the floor slab of the previous building structure example but no site activity can be performed without it. The cost of the site safety fence is an example of Project O/H that will be added to other costs of project activities. Among several examples, access and site internal roads, temporary site installations, utilities and services, site safety and security, storage and staging areas, and site horizontal and vertical transportation are very common ones. In addition, there are several technical and administrative services that a project may require which cannot be specifically charged to projects tasks. Costs of construction and MEP superintendents, designers preparing construction and shop drawings, as well as accounting, procurement, human resources, quality assurance, contract administration, planning, commissioning, training, safety-security personnel, and their offices are all project O/H examples.

The Project O/H is not a direct cost for an activity but is a direct cost of the project. Since it may/may not be practically possible to establish progress measuring metrics for such, one common way to recuperate Project O/H cost along with the project progress payments is by including it proportionally in the task costs.

Head Office O/H

Company head offices generate or develop the business for the project. Such effort may be a joint product of sales and marketing, estimate–bid preparation, legal, tax, finance, public relations, information technology, insurance–bonding staff and consultants. Furthermore, head office departments provide support to all site services previously mentioned on an as needed basis. These services include pre-contract, during project execution, project closure, and post-closure (i.e., guarantee period services and unsettled claim litigation) services. The Head Office O/H is generally distributed to projects on a project contract sum basis.

Estimating the O/H

The project O/H depends on project type, location, duration, and client's requirements. The Head Office O/H varies depending on company, operation size, and specifics of the service and/or product to be provided. Estimating O/H may be as cumbersome as listing all elements and costing them or it may be as simple as applying a percentage to the estimated cost; the latter being the least preferred but widely used practice. Although some of the O/H is variable costs, the generally accepted practice is to include them in the task costs as fixed costs:

- The Task O/H is itemized and included in the task cost estimate as a fixed indirect cost.
- The estimated Project O/H total is added to the project direct cost estimate (i.e., project O/H mark-up). That total is than distributed to each task as a percentage of the task cost.
- The Head Office yearly operating cost is to be recuperated by the projects actively generating revenue during the same time period. Since this cost is beyond the PM's responsibility, it is not included in the cost control process.

In Summary

- Both owners and contractors prepare cost estimates in the construction industry.
- The cost estimating process includes quantity take-offs for all components of cost line items.
- Cost estimate accuracies highly depend on the type of estimates and the amount of time allowed for preparing them.
- Expert opinion is required for specialty areas and where estimators do not have previous experience and/or where no credible reference is available.
- The engineer's cost estimate process starts with breaking down the scope to activities and ends with preparing the cost budget and preparing the expense and resource schedules after verification of the cost budget with the project schedule.
- The delivery method and contract pricing used for a project may shift the priorities of the stakeholders.
- The cost updated project work package dictionary can be sorted on a resource basis to generate project resource schedules.

7 Contingency as a Part of Cost

Introduction

Cost planning involves approximations, assumptions, and judgment calls when allocating resources and applying unit costs to project scope. When dealing with an approximated calculation of a probable cost based on gathered information, assumptions, and judgment calls, one should expect several types of risks. These risks stem from errors in gathered data, assumptions made, and various human errors as well as for expected but not estimated costs. Although some information may be available from prior similar work experience, by definition each and every project is unique and requires specific adaptation of that knowledge, if available. Contingencies are risk mitigation reserves. It is the most difficult cost component to calculate. It is the result of a judgment call, the accuracy of which depends on the expertise level of the person making that call. Using a contingency is a common and convenient way of handling uncertainties; it is an allowance for dealing with assumed errors and/or expected scope omissions. The definition of a cost contingency acknowledges assumed and expected risks. In estimating the cost of some task or an entire project, it is highly desirable to cover financial risks, but at the same time, overinflating cost which could potentially result in losing that business needs to be avoided. Contingencies can be used as a part of cost and as a part of the price markup. Contingency is an expression of comfort in an estimate or price. Contingency as a part of offered price—sometimes referred to as *management reserve*—is briefly discussed due to the merits of its use being dependent on the delivery method and contract pricing used.

Certainty, Uncertainty, Risk, and Contingency

Certainty is an ideal situation where all needed information is fully known or available for the user. To the contrary, *uncertainty* is a situation where there is absolute lack of information on outcomes, alternatives, and even their probabilities. Somewhere between certainty and uncertainty lies the *risk environment*. In this context, the source of risk is limited knowledge. Although there might be a great deal of knowledge on a specific subject, theoretically it is not possible to know everything 100%. Depending on many factors, the provided and/or available information will

DOI: 10.1201/9781003172710-7

contain some error which in most cases will be expressed as a plus or minus (i.e., ±) percentage. Error margins, confidence levels, and intervals and probabilities are examples of risk expressions when presenting information.

One way of working in a risk environment is by forecasting the risk. A *forecast* is the prediction of an event happening sometime in the future. Using models, a forecast produces a prediction of the likeliness of a certain occurrence. A common everyday example is the weather forecasts showing a percentage chance of an atmospheric event to take place in the future hours, days, and weeks. When the forecast is "60% chance of rain tomorrow" we understand that we had better take our umbrella with us since it is more likely that it will rain on that specific day. In this example we are given a 60% risk of rain the following day. That indicated probability, the likeliness of the subject event happening, expressed in a percentage or any other form, is a key indicator for the user.

An *estimate* is a product of assumptions and judgment calls. As discussed in the previous section, a *cost estimate* is an approximated calculation of a probable cost, based on available information, assumptions, and judgment calls. Several information sources, including forecasts, are used in cost estimates. Due to the fact that all information used is expected to have certain levels of accuracy and/or probabilities of occurrence likeliness, the calculated approximation of the cost using that information, the cost estimate, is inherently a risky result.

Since we know it is practically never possible to work in certainty, and because of the errors, tolerances, and probabilities pertinent to the information used in preparing cost estimates, using a **contingency is a partial mitigation of** that involved **risk**.

The Known and Unknown Risks

A project's risk environment can be analyzed in two sub-categories: the known and the unknown risks. The appropriate use of a contingency can only be justified for known risks. Since *risk is an unknown*, the project risk environment can be covered with these two sub-categories commonly referred to as *known-unknowns* and *unknown-unknowns*.

The Known-Unknowns

The risks in this sub-category emerge from adverse events that are expected to occur but it is not certain whether they will occur during the project and how adversely they will impact the project execution. Accidents on the job, equipment malfunctions and breakdown, and utility interruptions are examples of known risks. It is known—expected—that such adverse events may take place, but whether they will actually happen is "unknown". Furthermore, how big a damage, disruption, and/or any other negative impact such events will produce is unknown. Another known-unknown (i.e., risk) in cost planning is due to inherent errors of the cost estimating process. When preparing a cost estimate, we know that certain errors and omissions are likely. The main cause of such errors is the data used. The data used are expected to contain systematic, random, and other types of errors. Besides data errors, it is likely that by mistake, some parts of the job will not be included, due to various reasons including human errors. These are known sources of inaccuracies for any cost estimate. We *know* we are inaccurate but we do not know how inaccurate we are. Here the "known" is that we are inaccurate and the "unknown" is the magnitude of that inaccuracy. One other "known-unknown" case is where we know something is not complete, and/or erroneous but due to information and/or time limitations we cannot properly address the issue. Systematic and human errors are further examples for this category. We know these may happen and that the incidence will negatively impact our work, but we do not know when and how long and how damaging that impact will be.

A contingency allocated for some specific known-unknown should not have a cost significance. When there exists a significant known-unknown, a *provisional sum* may be the alternative to a contingency. A provisional sum provides a targeted not to exceed cost for a specific scope of work. A typical example of a provisional sum would be for the specific task of procuring and installing furniture. At the time when the building construction and fitting out contract is awarded, the owner may not be in a position to specify the furniture that will be installed but may have the budget figure for that work. By including this budget figure as a provisional sum, the guesswork and hence a contingency that may later be problematic is avoided for both parties.

Without any justification, a contingency of 5 to 10 % of the project cost is generally believed to sufficiently cover for known-unknowns of construction projects. It should be noted that this order of contingency is a total limit for all contingencies combined for project stakeholders and the stakeholders should clearly communicate their contingency considerations with each other since their included contingency will be compounded as it is carried over from one stakeholder to another. Considering a designer–builder–owner trio, if all of these three parties include a 10 % contingency to the same original cost, the final cost will include a 33.1% contingency instead of 10%. Although it is rare that all three stakeholders introduce contingencies for the same scope, duplications should be avoided by expressly communicating what has been included by whom.

The Unknown-Unknowns

This sub-category of uncertainties is about events that are unexpected and cannot be directly related to the project. In addition, the severity of their negative impact, if and when they occur, is not known. Such events are usually not accounted for since they are considered to be totally unrelated to the project. Still, when they occur, they may produce a direct or indirect negative impact to the project. Natural disasters, unusual accidents inside and outside the jobsite, acts of terror, riot, and strike are some examples for such unknown-unknown risks. It would be extremely difficult to predict if such events will occur during the project and to quantify their impact.

This is a category where justifying a specific contingency is difficult if not impossible. Instead of trying to cover an unknown with another unknown, a difficult to justify contingency, one should look for other instruments to cover such risks.

Major project contracts include "force majeure" articles for natural disasters and other major unknown-unknowns. A Force-Majeure article is intended to protect contractors from the risk of not being able to deliver contractual obligations due to such unknown-unknowns. In return, the contractors are expected not to include contingencies in their offered prices (i.e., not to inflate their prices) to mitigate risk from such adverse conditions.

Several types of insurances are available for stakeholders that cover unknown-unknown risks. When procuring insurance plans, a detailed coverage analysis will avoid duplications. In addition, special attention should be paid for not including contingency dollars to mitigate risks that are already covered in contracts and/or covered in some form of insurance of another stakeholder. Appropriately selected insurance coverage will provide greater cost and compensation certainty, especially when the resulting impact of such risks on the project is difficult to predict.

Contingency at Several WBS Levels

A contingency can be allocated for any task, at any level of the WBS. In addition, one could place a contingency for any cost component and for any line item under that cost component included in the cost analysis of that work. In theory, there is no limitation on where contingencies can be used. However, the amount of contingency that can be included in an estimate directly and conversely relates to the accuracy of that cost estimate. The more an estimate includes contingencies, the less accurate is that estimate. As a result, the decisions based on such high contingency estimates are less accurate decisions.

It is theoretically possible—but highly contestable—to place a contingency for each and every task and for every line item under the cost components (i.e., materials, labor, M/E, and O/H) for each task. Each one of these contingencies can be justified for one reason or another. It is assumed by definition that there will never be a 100% accuracy for any cost "estimate", but we would like to be as accurate as the case may require. The target here is to get as close as possible to that 100%. If contingencies were allocated for each and every level as previously mentioned, it is very likely that the targeted goal of 100% accuracy will be exceeded. Having too much contingency is equally as bad as having too little contingency. Allocating contingencies should be done for a specific reason and should never unnecessarily inflate the estimate.

Both Cost Elements May Include Contingencies

In addition to the possibility of placing them for tasks at several levels of a WBS, contingencies may also be included for both elements of task costs. The quantities in Table 6–3 are rounded off by the estimator to levels that he/she is comfortable with. Likewise, the production rates in Table 6–4 and the unit costs used in Table 6–6 are all adjusted/rounded off values that are judged to be the most appropriate ones for that specific task. That judgment call is an example of a contingency built in by the estimator that will be reflected in the overall accuracy of that estimate. Since these are *hidden contingencies*, it would be best declared and communicated with those whose contingency and business decisions will rely on such estimates.

Typical Uses of Contingencies

Contingencies are used to cover for errors made in cost estimating that include omissions to include some part of scope and for the adverse effects of expected events during project execution.

To Cover for Errors

Errors are integral parts of processes. They are inherent in the basic modern management functions in many types and forms. We know there could be many errors in our cost estimates due to wrong assumptions, measurements, calculations, and due to possible human errors including these and many similar other areas of human involvement. If we were to create an exhaustive list of possible errors in our project management processes, we would have to first accept the fact that due to some kind of an error we would never be able to make a 100% comprehensive list. Instead, general descriptions for some of the most common types are offered here.

Systematic Errors

This is a group of errors that occur due to inaccuracies of equipment used. Systematic errors will be repeated over and over again unless they are specifically addressed. Depending on the size of the operation, the magnitude of the error and its impact on the operation, the equipment and/or tools may need to be calibrated to a desired precision. The most common results of systematic errors are more or less usage of materials, longer or shorter labor and equipment hours, and larger or smaller utility (i.e., power, water, air, etc.) usage. If a measuring tool that measured a little more than what it should measure is used, more than needed of that material will be used because of that inaccuracy.

Random Errors

Random errors are harder to predict. We know they may happen but we also know they may not happen. We know there could be a power outage or some equipment could break down. Will that happen during our project, maybe. If and when such adversities happen, how big an impact they will have on our project is another challenge. Accidents can also be considered random errors.

To Cover for Omissions

When preparing a cost estimate, you can make deliberate omissions, or they may occur by mistake. Deliberately or by mistake, it is possible not to include some part of the work that is in the project scope. Likewise, while the work itself may be included, some related cost item may be left out by mistake or by choice. It will be the estimator's choice to make up for those omissions with contingencies. Other than the estimator, allocating some contingency for a scope omission and/or for cost line items is a difficult call to be made by anyone else.

When the estimator knows that a certain part of the work is not included, that estimator has two options to make up for that known omission. One option is to modify the related quantity and the second option is to adjust the unit cost (i.e., the cost elements) for the related work. What is important here is not to double up the contingency by adjusting both the quantity and the unit cost.

One other case of known omissions can be related to a long list of minor work line items that can be rounded off without a major impact on the overall cost. Rather than itemizing those line item costs, either a lump-sum amount or a percentage of the main cost items can be used to cover for the excluded long list of known minor costs. Seasoned estimators may have several best practices honed through experience that they can use to intentionally omit some work in both quantity take-offs and pricing. When preparing the quantity take-off for the plumbing system of a house, it is possible that the estimator could count each and every type and size of fitting that would be needed. Or, that estimator could allow a certain percentage of the cost of a certain diameter and type of pipe to cover for all the required fittings. Depending on the confidence that estimator has in that percentage, a contingency to allow for fittings and other consumables may be in order.

Escalation

Price fluctuations are a part of free market economy and so is *inflation*. Inflation can simply be defined as loss of buying power due to increasing prices. Especially for projects that are expected to take more than one year to complete, the effects of inflation needs to be factored in when preparing project costs. This is a known-unknown risk, especially for medium and long range projects. What is forecasted as price/cost adjustments due to increasing prices of materials, labor cost, energy prices, and any other cost item can be consolidated under the *escalation* title.

The escalation provision is a contingency to cover the effects of inflation and/ or for the changes of currency exchange rates of the owner and the contractor if these stakeholders are not using the same currency (i.e., the project is located in a country other than the owner's). Escalation is usually covered with a clause in the construction contract so that the contractor does not have to account for it with a contingency. Owners, on the other hand, have to include a contingency for escalation at the top level of their project cost budget.

Using a Provisional Sum to Avoid a Contingency

Contingency may not be a good option when dealing with specific scope exclusions and known-unknown scope changes. Such cases relate to a known scope exclusion with an unknown cost. In preparing the scope statement of a project, a major part of the project may not be ready to be priced out but it may not make business sense to hold back the entire project for that. In such cases, temporarily excluding that part of that scope with an order of magnitude cost may become necessary to prevent hard to quantify contingencies. As an example, when a mixed-use building development with a specific theme restaurant is ready to start construction, the specific equipment design of that restaurant may not be ready to be priced out. The owner has an order of magnitude estimate for that equipment and does not intend to delay going out for bids until such design is complete. If they are not given any design or direction, the bidders will be put in a position to include several related costs in their bids, such as shipping, handling, storage, and insurance, as well as all-inclusive bid and performance bonds. These costs depend on the overall cost of the subject unspecified equipment. Without any further information, the bidders will take their best guess as *contingencies* to cover for this known exclusion (i.e., unknown).

In this case, rather than delaying the bids for the project, offers from contractors can be obtained with a **"provisional sum"** for that specific exclusion for the restaurant equipment. When it is known that some work will be required but its scope and cost is not clear, a provisional sum can be allocated to cover the cost for that work if and when it is executed. This is a contingency for the known-unknown; the restaurant equipment scope being the known exclusion and its cost being the unknown.

With the provided provisional sum, contractors will have a better understanding of the magnitude of the undefined work, cap their contingencies based on that sum, and obtain indirect cost totals such as fees for insurances and bonds they are required to submit with their bids.

Dealing With Unknown-Unknown Scope Changes

A project may also encounter unknown-unknown work that needs to be executed but not included in the scope due to many possible reasons. An example for such unknown-unknown work would be differing—or unforeseen—site conditions that may not be included/disclosed in the bid and the contract documents. A very common example is jobsite soil conditions. Owners usually include soil reports providing detailed information on layers of existing soil types to be expected on the jobsite in the bid documents for potential contractors. A contractor uses that information to prepare its estimate for the stipulated earthwork. After contract award, during the excavation, if some rock is encountered where clay was reported, this is an example of differing site conditions which is an unknown-unknown scope change case. The contractor was provided with some information which did not include the type of rock that needs to be removed or the quantity of such work which is potentially a more costly and duration wise longer activity. Such unexpected (i.e., unknown) scope changes are best dealt with in a change order and contractors cannot win bids with arbitrary contingencies included to cover for such possible unknown work. Covering for the cost and duration consequences for such unknown-unknown work should be one of the considerations in determining the owner's general contingency for construction.

Differing/unforeseen site conditions are examples for situations where the unknown-unknown risk is a product of a series of errors and assumptions. This is not an issue that can be covered by insurance or by the contractor's contingency. Depending on the delivery type, the G/C may be well covered for adverse site conditions without a contingency but that may not be the case for the owner. The owner needs to have a contingency to mitigate such construction inherent risks that may pose significant cost and time impact on the project.

Contingency and Profit

One key factor for accurately calculating cost is properly dealing with contingencies. It is possible to use a contingency under any name or form and at any WBS level to minimize uncovered costs and dissipate uncertainties. If the overall project contingency is too little, it will lead to an underestimated cost and possible financial loss. On the other hand, too much contingency will cause overpriced offers and consequences for owners and other stakeholders. For the owner, an overestimated construction cost may negatively impact the project feasibility which may result in abandoning the project. For other stakeholders, and especially for the contractors, it may cause lost bids or business opportunities.

Included contingency is an indication of level of comfort, the confidence and risk perception of the estimator for the entire job, stakeholders, and key participants.

Some estimators may think that more contingency will lead to less risk and increased profits. Actually, the reverse of that is also a valid argument. If a party includes too much contingency in their bid for a job, they may never see any profit—but will incur the cost for placing an offer—if that party is not granted the job due to their inflated bid on that job. Unlike the rest of the cost components being easily quantifiable, contingencies are mostly unquantifiable, making them the most difficult cost item to judge and include.

Although impossible to entirely eliminate, the overall impact of contingencies may be minimized by higher precision quantity surveying and by thorough coverage of the scope of works. In building construction projects, cost estimates include contingencies ranging from 100% to 10% of the estimated cost, depending on the stage of the project at which that cost estimate is prepared. At the very beginning when the project is conceptualized, the cost estimate may use even more than 100% contingency. This is the "wish list" estimate. At the schematic design phase, when at least the size and shape of the building is more or less taking shape, a parametric estimate based on a square foot rate may have 30–40% contingency. During the design development stage, the project cost estimate based on a preliminary line item estimate is allowed to have a contingency of 20–30%. The final line item estimate (i.e., the engineer's estimate) usually includes a contingency in the order of 10%. The only time the cost for a project with no contingency can be summed up is at the completion of that project, and that would be the *cost of the project* and not an estimate anymore.

Theoretically contingencies for all cost components can be used, although it is unusual to have one for the overhead component.

Earlier the price was defined as:

Price = Cost + Profit Margin

That definition of price will be the following with the inclusion of contingency:

Price = Cost + Contingencies + Profit

and from the two definitions above:

Profit Margin = Profit + Contingencies

By this definition, the expected *net* profit depends on the use of included contingency; assuming that the contract will have a fixed price, less contingency use will increase profits. One could read this definition in different ways. One way of wrongly interpreting this relationship is to conclude that too much contingency cannot hurt your profits. After all, if the inflated contingency is not used, it will turn into a profit. This may be true if the contract for that project is granted approval from the stakeholders, or if this project was not outbid by some other contractor's proposal.

One other profit–contingency consideration is about how to calculate the expected profit margin for a project. From the G/C's perspective, the expected profit can be a fixed amount for a well-defined scope over a predetermined duration. If the project is to build a house with a complete design, with a well-defined scope, then a fixed profit will be a good option for a contractor to take on that job. In this case, the contractor is not in a position to allocate contingencies for scope changes and omissions do not reduce profit; the profit is secured.

The profit can also be a percentage of the cost. If an owner wants to use a certain contractor to build a house due to quality qualifications and for the expertise offered during developing the design, the owner may choose to pay a percentage of the cost as a profit margin (i.e., fee) to the contractor. Especially for older building renovation projects where several surprises and on the spot design solutions are frequently needed, this may be the fair approach for all stakeholders by eliminating guesstimated contingencies.

For projects that have substantial unknowns, utilizing delivery methods such as *cost plus fixed fee* is an advisable fair practice. If the profit were to be a percentage of the cost and an excessive contingency is included in the cost estimate, then there will be a concern in dealing with that excessive contingency since less cost will result in less profit for the contractor. In projects where the estimated cost includes a significant contingency for unknown-unknowns, a fixed fee which covers the contractors' general expenses and provides for some profit coupled with some cost saving sharing incentive for the contractor is a fair and preferred practice.

Contingency and Change Orders

Change orders are known-unknowns of projects. Every project encounters a change of some sort that causes deviation from the original scope, cost, and duration. Obviously using a contingency to cover the cost impact of an unknown change is an effort in vain. Changes with minor cost impacts such as material type changes for a similar material and small quantity variations can be accommodated with only adjusting the contract sum. Larger changes that also impact the contract duration may require a more comprehensive price adjustment since it may involve additional head office and project O/H.

While the owner may use some contingency to cover for scope changes along the project duration, using contingencies to deal with possible future change orders is not a good practice for contractors and should be avoided. Instead, pre-agreed unit prices for resources can be included in the contract documents which can be used to determine the cost of the issued change order if and when one takes place.

Hidden Contingencies

Estimates include several hidden contingencies. Contingencies are not necessarily dollar sums; they can be included in either cost element. Rounding up a measured quantity is one way of placing what may seem like a small contingency in that quantity. If a unit cost that has also been rounded up to a higher value is used to calculate the cost of that work with the rounded up measurement, the resulting contingency is compounded.

In a simple example of calculating the material cost of a run of cable:

- Measured length: 87 yards; rounded up to 100 yards (because in the absence of a measuring tape, you have chosen to count your steps)
- Unit cost $ 3.55 / yard; rounded up to $ 4 / yard (since you have taken this information over the phone and feel like it should be a little higher

Will result in calculating the material cost as $ 400 instead of $ 308.85 per run. That is more than 30% contingency for this item.

Depending on numerous factors such a result could be considered extremely misleading, or perfectly all right. While choosing to use a safe quantity and unit cost could be prudent, overdoing it and not disclosing such in documentation could be disastrous. If and when the inclusions of contingencies are not disclosed in communicating and exchanging information with team members, such hidden contingencies may cause major errors, misleading the upper management.

How Much Is Enough?

Contingency is the most difficult cost to quantify. Although eventually it will be converted to a dollar amount, a contingency can be in the form of time or as a quantity or unit cost for any cost component line item. A good business practice should not allow profits to depend on uncertainties. Once the primary costs and risks are known, the profit to be made from that project should be determined independent of any uncertainty that may originate from scope changes, delays, unexpected/differing conditions (i.e., undisclosed or wrongly disclosed information in bid documents), and other similar causes encountered during the project.

Contingency is not a remedy to risk. It is an adjustment for allowable errors. In that respect the total amount of allowable contingency can be related to the expected precision level of the cost estimate being made. Contingencies should not be the preferred tool to cover the expected inaccuracies for Order of Magnitude and Parametric cost estimates. For initial and final line item cost estimates, where the expected accuracy is between 10% to 25%, the allocated contingencies should be less than or within those accuracies and never exceed them.

Since a project may involve several different entities as stakeholders, each one of those stakeholders may have a wide spectrum of their own risk and comfort level assessments. From the perspective of contingency being a security reserve, a reserve alleviating the perceived project risks of stakeholders, determining a decisive contingency magnitude for every individual project stakeholder depends on factors well beyond the fundamental ones discussed in this chapter.

Management Reserve and Project Manager

The stakeholders of a project may choose to place some allowance in their costs/ prices and call it the "Management Reserve". This is basically a "contingency" for that stakeholder's own use and purposes. Depending on the delivery method and contract pricing used, such contingencies may/may not even be disclosed to other project stakeholders. While extensive definitions and rules are out there to consider, the simplest rules for using such contingencies can be highlighted as:

- If a contingency is there, it needs to be there for a reason and it needs to be used for that reason, if and when the underlying condition of that reason materializes.
- The PM has the authority to use contingencies approved in project plans for their designated purposes.
- In cases where there is unused contingency for a specific use and there is a need for additional resources for a different use, the PM may be allowed to make that shift or, he or she may be required to notify higher management and ask for a plan change approval.
- Using time contingencies that impact project milestones and completion dates need special attention and they need to be formally communicated with other project stakeholders as soon as practically possible.

In Summary

- A contingency is an allowance for errors, overlooked parts of project scope, and risks.
- Contingencies partially mitigate known risks.
- Considering contingencies for unknown risks will unnecessarily inflate estimated costs or prices.
- Contingencies are the hardest cost to judge and/or calculate, so they should be allocated with utmost care.
- Theoretically every cost element and component of an activity may be allocated a contingency.
- Project planning may include a contingency in any form of resource such as time and material as well as money.
- Contingencies can be allocated with top down or bottom up approaches.
- Overuse of contingency does not ensure higher profits; most probably it will overinflate the price and result in losing contracts.
- Total contingency weight (i.e., percentage) should not exceed the expected accuracy of the estimating method used (i.e., no more than 10% overall contingency for an engineer's line item estimate).
- Total included contingency should not exceed the profit margin.
- Contingencies cannot be billed on their own; they need to be blended into billable contract prices.
- Appropriately communicating included contingencies within every stakeholder organization is crucial in eliminating excessiveness and duplications.
- The project managers of the stakeholders have the authority to use included activity and/or project contingencies.

8 Project Cost Budget

Introduction

One of the most common depictions of project cost comes in the form of a budget. A budget is a plan with resources allocated to tasks. The cost budget for a project is a plan with resources allocated for performing the tasks of that project.

Budgets are generally presented in terms of money, but better prepared budgets include more than only dollar amounts; they provide the details for allocated resources, which is vital information for project managers. Costs estimated for the tasks defined by the WBS provide the basis of a structured project cost budget that can be presented at any required detail level. The budget is a performance reference for scope, cost, and time constraints. The summation of specific component line item quantities provides the planned project totals for specific resources; the resource budgets. Resource quantity-based budgets for specific critical tasks of projects are preferred over dollar amount budgets since they provide better references for control purposes. Budgets can be prepared in many formats to serve the needs of managers at different levels and for reporting to stakeholders. Depending on the format used, budgets can be used to monitor the past and present and to project the expected future performance of a resource, task, and the project. Predetermining criteria for when to make changes/revisions is a well-accepted practice. For projects that generate revenue, project cash flow is created by spreading both expenses and revenues over their timelines. Cash flow analysis is essential for determining if, when, and how much financial support a project may require, especially during the construction mobilization phase when mostly unbillable indirect expenses are incurred by the contractors.

Project and Operational Budgets

Project budgets differ from operational budgets in a number of ways. Project budgets cover the entire project duration and they require unique resources.

One-Time Occurring and Not Repetitive

A project budget is a one-time budget since the project is a one-time event. This represents a difference between a project budget and a periodically repetitive

DOI: 10.1201/9781003172710-8

operational (i.e., program) budget. The project budgets require a more proactive control since some activities may take place only once or the flow of the activities may not allow for corrective action; once the resources are consumed there may not be a second chance for remedy.

Covers the Entire Project Duration

A project budget covers the costs of the project throughout the project duration. It is not prepared for a specific period (i.e., week, month, or a fiscal period) which would be the case for operational budgets. Since operational budgets are periodically repetitive, the information collected by monitoring the previous budgets provides substantial reference and they can be used in preparing the following one. Such reference information is rarely available due to the uniqueness of every single project. The estimators and managers working on the costs and the budget need to make judgment calls to make adjustments.

Requires Unique and Project-Specific Resources

Unless it is the very first time, the budget performance of prior periods (i.e., years) is available when preparing an operational budget. Most labor, M/E, and indirect resources are already engaged and are a part of the operation. For a project, it is quite the opposite. Every resource is to be procured, mobilized, and put to work as per the specifics of the project's one-time operation. The data used and assumptions made during preparing the project budget serve as the basis for procuring the project resources whereas an operational budget is based on existing resources with minor adjustments as may be needed.

Project Cost Budget Includes and Reveals All Required Resources

The project cost budget outlines what resources are estimated to complete the project and establishes performance metrics to measure accomplishments during execution. As is the case for any cost, the budget is NOT only a list of expenses and revenues of dollar amounts, although dollar amounts are the most visible on the front page. The estimated quantities, unit costs of resources, and durations for activities are equally important parts of a budget. Since the budget dollar amounts are the products of quantity and unit cost elements, the underlying assumptions (i.e., estimated cost elements) are crucial pieces of information for the project manager, especially for controlling project costs. Although only the resulting dollar amounts are commonly used for presentation purposes, without the supporting data on elements and assumptions, a budget is nothing more than an unmanageable wish list.

Project Cost Budgets and Schedules Are Interrelated

Although project cost can be obtained by summing up the costs in the WPD, the project duration cannot be calculated by simply adding up the durations for those tasks. Project duration is determined by scheduling those tasks. Because of their interdependency, determining project cost and duration is expected to be an iterative process. Once one is known the other can be checked and verified and vice versa. Without a definitive project duration, the initial WPD may not accurately include the costs for some duration-dependent indirect costs. However, it is extremely rare that a project's duration will depend on some indirect cost activities. As an example, without knowing the project duration, the final cost for some rented site installations cannot be determined. The duration of site installations' lease will not be a factor in determining project duration. For that reason, once all the direct costs of activities and their durations are determined, they can be scheduled to determine the project duration. With that project duration, the time-dependent indirect costs can be determined (i.e., the duration hence cost for leased site installations) and/or updated to be included in the overall project cost and budget. At the initial stages of a project when a detailed schedule may not be available, budget preparers may have to assume a project duration and estimate those duration-dependent costs accordingly, knowing that they may have to revisit these specific costs when a more precise project duration is available to them.

Project Budgets Include All Known-Unknowns

Project budgets include all knowns as well as known-unknowns. Including known-unknowns may require judgment calls and assumptions. Including assumptions in the WPD will ease updating them when more information is available. In addition, this structured approach will ease the communication of future revisions with involved staff and stakeholders. Unstructured and undocumented assumptions may necessitate budget revisions which may raise questions and may cause contention among the stakeholders. Exclusions should be used in agreement with stakeholders and staff. Even with detailed specifics, excluding parts of project scope should be the last option when preparing budgets.

Every project budget should have a safety net, a general contingency, to tap into if and when needed. The big question is *how big* this safety net should be. There is no rule and/or best practice to use when determining how big a contingency a project will require. The decision on the magnitude of a general contingency to cover the project in its entirety will depend on many factors, as discussed in the previous chapter.

Preparing and Reporting Cost Budgets

Preparing the budget starts with converting the costs to budgets. Using a bottom-up approach, first the activity budgets will be prepared which then will be summed up to form the project budget.

Budgeting Activity Costs

Usually the cost information in the WPD does not include the final budget number for tasks. The main difference between an activity cost and its budget figure is the O/H and contingencies not being included in the cost. An activity cost may include a specific indirect cost for that activity which is the task O/H. The budget preparer will have to consider including any other indirect costs and/or contingencies for specific tasks. Using the estimated activity cost of Table 6-6, those considerations and the process is summarized in Table 8–1.

The following highlights the budget preparer's considerations for this sample task's budget:

- The $750 O/H is a task-specific indirect cost (see Table 6–6 for details)
- No additional O/H (i.e., project or head office O/H) is considered
- Component contingencies specifically for materials and labor components have been added
- In addition to the component contingencies, a task contingency has been included

The component and task contingency numbers given in this example are demonstrative numbers and they are not to be construed as necessary and/or as indicative in magnitude.

Once the task budget is prepared, it should be included in the WPD together with the task's cost estimate.

Budgeting Project Cost

Project budget is the summation of all task budgets, all project and head office overheads, and contingencies for all sections of the job and at all levels of the WBS. In its full detailed form, it includes all line items for activity cost elements. Due to the lengthy details involved, listing only activity budgets will suffice for middle to upper management levels. However, the full detailed version for sections of the job is essential for those who would be periodically monitoring these activities in their respective sub-divisions.

The basis of the project cost budget is the WBS activities with assigned budget values to them. Similar to the indirect cost and general contingency considerations for tasks, the budget preparer will include the general indirect costs and

Table 8–1 Example for preparing a cost budget for an activity

Description	3.2.4.2.1. Floor Slab Cost Budget			
	Materials	*Labor*	*MEOD*	*O/H*
Estimated cost	117,670	87,060	1,000	750
Component contingency	7,330	2,940	–	–
Subtotals	125,000	90,000	1,000	750
Task subtotal	216,750			
Task contingency	8,250			
TOTAL	**225,000**			

evaluate the need for sectional and overall contingencies before finalizing the project cost budget. The general indirect costs at this point are the project (i.e., job site) general costs and the portion of the head office cost that will be charged to this project.

Budgeting for Project Resources

Similar to what is given in Figure 6–6 for activity costs, the monetary project budget can be regrouped and/or compiled to provide the quantities for project resources. Figure 8–1 shows how activities can be listed to reveal the quantities for their material components. Similarly, the project quantities for other resources can be derived from task budget data by tabulating the labor, M/E and O/H quantities of project tasks.

Based on their significance and practicality, only critical resources need to be resource budgeted. Determining which resources are critical will depend on the tasks they relate to as will be further detailed in Chapter 9. In calculating the project resource quantities, the contingencies included in the task budgets should also be included. Unless specifically allocated for a specific resource of a task, contingencies can be included by using a multiplier which would be the ratio of the contingency to the cost total.

Budget Formats

Depending on its user, a budget can include different details and WBS levels. A budget that contains the full WBS task level details may not be suitable for all stakeholders and for all management levels. Budgets can also be used in parts; rather than the entire project budget, specific parts of the WPD would be more helpful for a first line manager in charge of a specific section of the project. On the other hand, a top-level manager will be more interested in seeing the entire project status at one glance and would prefer a less detailed—a top WBS level—budget.

Several budget formats can be used depending on:

- Project complexity and duration
- The phase and estimate precision of the project
- Stakeholder being reported
- Reporting past and present periods or projecting for project completion

Monitoring Cost Budgets

Project cost budgets are monitored and reported periodically. The monitoring/ reporting period is commonly a calendar month for construction projects. However, for their own internal use, mid to lower level managers can monitor cost/ resource budgets and issue reports more frequently as they may find necessary to monitor their teams' performances.

Activity		Material quantities					
WBS #	name	Material A	Material B	Material C	Material D	All other material	Last material type
....1	first task	1st task's Σ for Material A	1st task's Σ for Material B	1st task's Σ for Material C –			
....	...	...					
....	...	...					
....	...	...					
....last	last task	last task's Σ for Material A	last task's Σ for Material B	last task's Σ for Material C –			
Project totals –		Σ material A	Σ material B	Σ material C	Σ material D	Σs for all others	Σ last material

Figure 8–1 Calculating total material quantities for the project from task budget data. Other resource totals for labor and M/E components can also be obtained in a similar way as the figure demonstrates.

Budget Format for Performance Monitoring

Figure 8–2 shows a preferred format for monitoring past and current budget progress. It tabulates the current total project budget for a line item, the budgeted and actual incurred expense of the reporting period, and the cumulative of the budgeted and incurred expenses up to the reporting period, side by side. It also shows the periodic and cumulative deviations from the planned targets.

This is a simple performance monitoring format that displays historic and current accomplishment data, especially useful for performance evaluations for selected critical activities. The performance data obtained from this format may serve better for low to mid-level managers, who will be more interested in improving critical activity performance.

This format can also be used to monitor/report planned and achieved revenues.

Tabulating the cost budget values for "reporting period" against "cumulative to date" in this format indicates that the line item totals of this budget have to be distributed over their scheduled durations. This coupling of budget and scheduling is easily accomplished if they are both based on the same WBS activities and enables the managers to monitor monetary, physical, and time progress of project tasks easily and consistently.

Budget Format for Futuristic Projections

Figure 8–3 shows a budget format which incorporates futuristic projections. This format reports the actuals including the committed costs and uses that data to project the remaining activity cost and the total cost of the activity at its completion. The futuristic view this format presents makes it more useful for middle to upper managers who are more proactively interested in knowing the challenging activities and how well the cost budget will do at project completion.

Since the column headings are abbreviated in Figure 8–3, the next paragraphs will provide details.

The first three columns of this format define the WBS number, name, and initial budget value for the reported tasks.

Activity	Project Budget	REPORTING PERIOD			CUMULATIVE TO DATE		
		budgeted	actual	variance	budgeted	actual	variance
Total							

Figure 8–2 Simple budget monitoring format.

WBS Number	Description	Initial Budget	Approved Changes	Adjusted Budget	Committed Costs	Pending Costs	Cost to Complete	Cost @ Completion	Over/ (Under)
TOTAL									

Figure 8–3 Budget reporting format with projections.

The "Adjusted Budget" column indicates the latest (i.e., current) budget value if the corresponding task's initial budget value has been revised. If not revised, this column will show the initial budget value or, as an option it can stay blank until it is revised.

"Committed Costs" refers to the portion of work that has been either completed, procured, employed, and/or contracted. It indicates that there is some sort of an achievement and/or commitment obtained for the amount being reported. It does not indicate completion, delivery, and/or payment, although some completion, partial delivery, and payment might have taken place. Committed Cost only indicates that the portion of the cost of that line item is secured and there will be no change to that indicated cost until project completion.

For example: If a purchase agreement is made to procure 10,000 cubic yards of sand at a unit cost of $35.00 per cubic yard including tax and delivery, the committed cost for "sand" is $350,000 although none of that sand might have been delivered at that point and regardless of any payment made or not. During the project execution, when some quantity of sand has been delivered and some payments have been made to this supplier, the committed cost total of $350,000 will not change regardless of what has been delivered and paid for.

For the same example if there was **no** purchase agreement and 2,000 cubic yards of sand was procured at the all-inclusive unit cost of $35.00 per cubic yard, regardless of how much was actually paid, the committed cost would be $70,000.

"Pending Costs" refer to the remaining budget figure after deducting the "Committed Cost" from the "Adjusted Budget." For the specified task or task line item, the "Pending Cost" is what the budget allows for further committing at the time of reporting.

Using the same "sand" example, if the "Adjusted Budget" was $450,000 and the "Committed Cost" was $350,000 for this line item, then its "Pending Cost" would be $100,000 regardless of received quantity of sand and the amount that has been paid/unpaid at the time of reporting.

Pending Cost does NOT indicate that the specific work can be completed for this amount.

"Cost to Complete" (CTC) is the cost that the PM projects and/or estimates as what it will take to complete that work *from the reporting point forward*. CTC is calculated by using updated and/or projected quantities and unit costs that the PM observes as applicable to the cost elements of the subject task line item. In other words, CTC is the product of the remaining quantity of work at the going rates. It is worth noting that the remaining quantity of work may differ from the initially estimated quantities, which on its own positively or negatively impacts the final cost. Naturally, any change for the unit costs will also directly affect the final costs.

CTC is calculated by using information gathered from a number of sources. The cost data on elements of what has been accomplished, quantity of remaining work, efficiency trends of teams, changes expected in site/weather conditions, and market trends are some of the things that need to be looked into when calculating the CTC.

"Cost at Completion" (CAC) is the projected cost for any budget line item at the completion of the project. It is determined by first calculating/estimating/projecting the "CTC" and adding that to the Committed Costs.

By processing the Adjusted Budget and CAC, the status of that specific work item is reported in the "Over/(Under)" column. When the "Pending Cost" equals the CTC, the "Over/(Under)" column will be "0"; indicating no variance to current budget is expected till completion. Another column can be added to this format that would indicate the Over/(Under) column as a percentage, especially if the budget revision thresholds are established with percentages as well.

Budgets reported with the format of Figure 8–3 provide a more comprehensive status of the complete project from a cost perspective. Appendix A provides several stakeholders' sample project cost budgets in various formats which serve best for their purposes.

Monitoring Project Resources and Output Rates

A project may require all sorts of resources, money being only one of them. Monetary budget reporting does not reveal the actual physical work accomplished although most commonly budgets are prepared in terms of money. Monetary budget monitoring only reveals the committed costs and the remaining costs to complete the project. Actual physical work achieved needs to be measured on the terms of its cost elements, quantities, and resources.

Measuring accomplished physical work is done by monitoring both physical quantity of work completed and resources used in the execution. QTO provides the physical quantities of works to be accomplished and the actual physical progress is obtained by simply measuring the actual work put in place. During the initial planning of the project, the resources that will be required to accomplish the QTO quantities of work are calculated by planners. That calculation is based on assumed output rates for labor and M/E resources and that assumption is one key factor in determining the initial costs and durations of activities. When the project is in motion, all of the *assumed* quantities and output rates need to be monitored and verified for the physical work accomplished for critical activities.

Resource-based budgets are essential tools to monitor the progress of actual physical work output. Rather than using the costs (i.e., product of resource quantities and unit costs), the resource budgets are prepared by sorting the quantities of the same resources required for activities. Monitoring the budgeted resources for critical activities based on accomplished performance will provide the details of a specific resource's performance and/or its sufficiency when compared to its estimate. Similar to CTC and CAC, the PM can project resources to completion and at completion for critical tasks.

Generally, resource-based monitoring is required only for critical project activities. Monitoring only monetary progress for these activities may not provide sufficient information on their current and/or future status. Combined monitoring of physical progress and resources used to accomplish that physical progress provides a more complete status report for subject-critical activities since they will reveal the status of planned versus achieved output rates that were used in estimating the costs and/or estimating the durations of those activities.

Figures 8–4 and 8–5 will be used to demonstrate how monetary and resource budgets can reveal differing conclusions. The labor component of the project depicted in Figure 8–4 is selected as a critical task for a project. Figure 8–4 shows monetary budget reporting of that project at the end of the first period (i.e., the first month) and Figure 8–5 reports the cost elements (i.e., the resources) of the labor cost for the same reporting period.

The reporting displayed in both of these figures may be construed to indicate that the project is right on track since the reporting shows no variance to what was budgeted. However, a close look into the cost elements of the critical "labor" component, as displayed in Figure 8–5, presents a different view.

On the "BUDGET" side, the elements of the labor component shows 1,000 labor-hours, at a unit cost of $ 100/hour, resulting in $ 100,000 as the monetary

Expenses	Project Budget ($)	REPORTING PERIOD ($)			CUMULATIVE TO DATE ($)		
		budgeted	actual	variance	budgeted	actual	variance
Materials	330,000	110,000	110,000	0	110,000	110,000	0
Labor	*300,000*	*100,000*	*100,000*	*0*	*100,000*	*100,000*	*0*
M/E	24,000	8,000	8,000	0	8,000	8,000	0
Overhead	120,000	40,000	40,000	0	40,000	40,000	0
Contingency	26,000	–	–	0	–	–	0
Total	**800,000**	**258,000**	**258,000**	**0**	**258,000**	**258,000**	**0**

Figure 8–4 Budget monetary progress reported at the end of its first period for a sample project. Highlighting and italics indicates the labor component is critical for this project and will be further monitored on the basis of its cost elements.

DESCRIPTION	REPORTING PERIOD						
	BUDGET			ACTUAL			
	quantity	unit cost	total cost	quantity	unit cost	total cost	variance
Labor (selected critical task)	1,000 hrs.	$100/ hr	$100,000	1,250 hrs.	$ 80/ hr	100,000	0

Figure 8–5 Resource budget reporting for the critical labor activity of Figure 8–4 sample project. The selected activity comprises only one resource, only one labor trade, for simplicity's sake of this example. Resource budget format used can also include "cumulative to date" columns to reveal the overall achieved costs element values of previous periods.

budget for the reporting period. Although the reported "ACTUAL" reaches the same total of $ 100,000, the labor hours shows an increase, from 1,000 hours to 1,250 hours, and the unit cost shows a decrease from $ 100/hr. to $80/hr. If the reduced hourly rate is sustainable throughout the project, that could be considered advantageous at first glance. But coupled with the increased hours used, that should raise a red flag:

- Why is the estimated production (i.e., output) rate not achieved?
- Is the lower hourly rate related to the reduced production rate?
- Will the additional hours needed cause an extended task duration and will it cause project completion delays?
- If there will be a delay, how will that impact the project O/H (i.e., delay penalties)?

These are some of the questions that need to be answered and they will be a part of the PM's controlling function, which will be discussed in further detail in the next chapter.

Planning project resources and monitoring their progress on their cost elements is an essential part of project management. All direct cost components, materials, labor, machinery and equipment, contain potentially critical resources. Selecting the critical tasks for the project and detailing their most prevalent resources, the critical resources budget of the project is an essential management tool that reveals the depths of project progress status. Resource budgets, unlike the monetary ones being only in dollars, are presented in the specific units those resources will be measured by and will indicate the quantities of the resources planned to be consumed in order to accomplish the respective project tasks.

Labor and M/E resource budgets are the most common resource budgets used to monitor efficiencies for these resources. They are prepared in labor-hours and machinery-hours for different labor trades and machinery types. When evaluated together with monetary budgets and physical work they were estimated to produce, they will provide invaluable performance insights to PMs.

The output rates of labor and M/E obtained by monitoring actual performance on site is an excellent reference since these elements are less location and inflation sensitive than unit costs and hence more transferrable to other projects.

Figure 8–6 shows the estimated resources of an activity on a past and current performance budget format. Obviously, all of the line items depicted there cannot be critical. The project manager will select which line items need to be monitored and will request site feedback on those items. With that, the reporting site managers will not be bogged down with unnecessary and time-consuming details and will focus on the critical ones.

Revising Budgets

Budgets, like any other plans, are expected to change. When there is evidence that it is not working as expected, what is not working in the plan needs to be

| DESCRIPTION | REPORTING MONTH | | | | | | | CUMULATIVE AS OF END OF REPORTING MONTH | | | | | | | | REMARKS |
| | BUDGET | | | ACTUALS | | | | BUDGET | | | ACTUALS | | | | |
	qty.	unit cost	Cost	qty.	unit cost	Cost	variance	qty.	unit cost	Cost	qty.	Unit cost	Cost	variance	
3.2.4.2.1 FLOOR SLAB															
MATERIALS															
Ready mix concrete incl. transprt. & pump.	273	175	47,775			–									
Forms															
Metal props for deck (100 uses expected, 1/100 cost charged)	700	2	1,400												
4x4"x8' beams (10 uses)	700	2	1,400												
3/4" plywood (5 uses)	480	8	3,840												
2x4"x8' wood (10 uses)	300	1	360												
Wood planks 2x8"x10' (10 uses)	100	1	50												
Pegs 1x2x12"	1	0	20												
Form oil	100	6	600												
Nails, screws, tie rods, braces, and other consumables.	1	400	400												
Reinforcement															
Rebar delivered cut, bent & tagged	18	3,000	54,600												
Rebar wire	500	1	625												
Spacers	400	5	2,000												
Miscellaneous															
MEP	1	2,500	2,500												
Anchors	1	1,500	1,500												
Plastic moisture barrier	14,000	0.100	1,400												
Burlap for water curing	14,000	0.200	2,800												

Figure 8-6 Budget progress monitoring format for example floor slab task of Chapter 6. Not all of these line items need to be monitored as will be explained in Chapter 9.

DESCRIPTION	REPORTING MONTH							CUMULATIVE AS OF END OF REPORTING MONTH							REMARKS
	BUDGET			ACTUALS				BUDGET			ACTUALS				
	qty.	unit cost	Cost	qty.	unit cost	Cost	variance	qty.	unit cost	Cost	qty.	Unit cost	Cost	variance	
Water for curing	100,000	0.004	400												
Total Materials			117,670												
LABOR															
General Foremen	110	60	6,600												
Carpenter	766	40	30,640												
Iron worker	100	40	4,000												
Laborer	1,732	25	43,300												
Electrician	24	45	1,080												
Plumber	32	45	1,440												
Labor Total			87,060												
M/E			—												
The site installations, storage/ staging areas, vertical and horizontal transportation equipment and similar other resources is used but their cost is not included. The general site opertaion cost will be calculated and added as a general overhead. Only specific handtools, lighting, power extensions and other similar tools/ equipment provided are included here.	1	1,000	1,000												
M/E Total			1,000												
O/H			—												
Only task specific items such as specific gear, cleaning and repair supplies are included here.	1	750	750												
Total Overheads			750												
Overall Total			206,480												

Figure 8-6 (Continued)

changed. Changing a plan is not only required when things are not working as expected. The opposite is also a valid argument for revising plans. *Continuous Improvement* is the never-ending management process of looking for performance and quality improvements. The possibility for some improvements may as well be the reason for revising budgets.

Before going into any further details of revising a budget, a distinction between *updating* and revising needs to be made. A revision is a significant change to the existing budget and requires approval(s). An update does not require any approval since it does not present a significant change. It only renews, complements, and/ or supplements the existing with newer and improved data. It is the process of incorporating periodically collected, sorted, and/or calculated cost data into the cost budget. Updating is an ongoing process whereas a revision takes place when there is a need, a significant deviation from what has been targeted.

From a scheduling perspective, a cost budget update may not result in a change for project duration. A cost budget revision is more likely to cause a project duration change. Either way, the schedulers have to be fully informed of all cost budget updates and revisions for their further action.

Cost budget updates and revisions mandate revisiting expense and resource schedules and incorporating the updated/revised data into them.

In revising budgets, timing is critical. Naturally, all sorts of changes are expected to take place when a project is in motion. All cost elements for all components may show positive and/or negative variances. Does the PM revise the budget each time he/she reports those changes? The answer is no. The budget revisions should only take place when there is a *major* change and whenever that occurs. Defining what is considered a *major change* is the key argument in budget revision decisions.

Since every project is unique, there could not be a standard set of criteria defining what should be considered as a major change to one specific project. The goal here is to establish an early warning system so that potential budget overruns are managed in due time. Setting a variance threshold would be an option. If that threshold is set as a percentage, the budget will be revised when the task cost budget—budget excluding the project overall (i.e., blanket) contingency—variation is that much above or below the target. At this point, the project overall contingency may cover the variance and the project could still be considered within budget. However, even if the targeted values have not been exceeded, checking out the cost trends by calculating the CTC and CAC and updating the budget should be the preferred practice.

Another option would be setting a value, a dollar amount, as a threshold value for variance. If and when a task variance reaches and/or exceeds that value, that should be a good alert to run a CTC/CAC check for that task.

In any project the budget weights of tasks may vary substantially. Using only a percentage on the task variation or a fixed variance threshold amount applied to all tasks may not serve as good indicators on their own. A small percentage of a large critical sum task may well exceed a hefty portion of another task with a smaller sum. Likewise, a small change for a large task may result in a major change for the project. The best practice is to set threshold indicators as a percent variation of tasks and a not to exceed fixed amount, both at the same time and along with similar threshold values and percentages set for project cost budget total.

It is quite common that during a project, some task costs may show an increase, some others may show a decrease, and yet some task costs may hit the target. In the hypothetical case where all the positive and negative variances add up to zero, it would be wrong to conclude that there were "zero changes" to the budget. Compensating deficiencies with surpluses may disguise the root causes of problematic budget allowances for tasks. To address such situations, a variation threshold percentage based on the *absolute variations* is a good practice to call for a budget revision. The absolute variations is the total variations to the project budget which is determined by adding up the absolute values (i.e., all variations as positive numbers) of the variations.

Budget–Schedule Integration: Expense Flows and Resource Schedules

Unlike operating budgets for programs having yearly budgets, project budgets span the entire project duration. When the same WBS activities are used by cost and time planners, the costs of tasks can easily be spread over to their time slots in the project schedule.

Similarly, the revenues generated by tasks can be distributed over their scheduled durations. If the contract price is based on the same WBS, preparing the project cash flow becomes much easier since both cost budget (i.e., the expenses) and progress payments (i.e., revenues) are on the same task basis. Otherwise, calculating the revenue a specific task will generate may not be straightforward; tasks may need to be grouped/regrouped to calculate their revenues.

Preparing the CF requires a precise positioning of activity expenses and revenues in their respective timing slots in the project schedule. Figure 8–7 shows a typical cash flow for a project activity. The total cost for that activity is $ 350,000 but that cost is not totally incurred during the time that activity is shown to take place in the schedule. The cost components of this activity is tabulated in Figure 8–7 as their costs will be incurred in relation to the time period this activity is scheduled to be conducted. Some cost components have to occur before others, such as ordering and making materials available. Some cost line items of an activity (i.e., keeping safety and protective equipment in place) incur cost after completing the work, since they have to stay in place after the execution of the work is complete. The overhead cost is incurred as long as the project is not closed. Figure 8–7 shows a typical case of how these costs could incur before, during, and after an activity's scheduled duration.

As is the standard for construction projects, but depending on contract conditions, the related progress payment follows completion of work and does not take place at the exact time that costs are incurred, but it will be close. For the case depicted in Figure 8–7, the progress payment (i.e., revenue) for this activity is expected within the same time period (i.e., same week or month) when the costs

	task timeline (days, weeks, etc.)				
	preparation (before)		execution	maintenance (after)	
	-2	-1	0	1	2
task schedule			▆▆▆▆		

task budget		task budget spread over task timeline				
material expenses	$100	$70	$20	$10	–	–
labor expenses	$140	–	$60	$80	–	–
M/E expenses	$60	–	$20	$30	$10	–
O/H	$50	$10	$10	$10	$10	$10
expenses total	-$350	-$80	-$110	-$130	-$20	-$10
revenues	$400	–	–	–	–	$400

periodic cash flow		-80	-$110	-130	-$20	$390
cumulative cash flow		-$80	-$190	-$320	-$340	$50

Figure 8–7 Typical activity cash flow details. This figure shows expenses in a timeline being incurred before, during, and after the time when activity takes place on the schedule. Likewise, due to billing and processing, the revenue is expected at a later time, creating a cash flow deficit until it is cashed out.

are incurred. When an activity generates a revenue (i.e., a progress claim is made), it will take some time to process that claim before the actual payment is made. It takes about 30 to 60 days for most owners to make a payment to their contractors. This payment process duration is usually provided in the contract.

Another standard for construction contracts is *retention*. The retention is an amount withheld by the owner as a security/warranty for completed work. Commonly the retention is 10% of the activity price. When an activity is completed, the contractor makes the progress claim for 100% of the price but gets paid only 90%, the 10% being withheld as the retention. The common practice is to release all or part of the withheld retention at *substantial completion* of the project and replace it with a bond to cover for *latent defects* that may have to be remedied during the *guarantee period*, following the substantial completion. Any remaining portion of the withheld retention and/or related bonds are released at the end of the guarantee period, at project completion.

If the contract articulates retention, that withheld retention needs to be accounted for in project cashflows. The example shown in Figure 8–7 does not account for withheld retention, implying that there was no such contractual requirement for it.

Generating the budget–schedule—and the revenue stream—integration enables the PM to prepare the financing his/her project would require. Breaking down the expenses to their components further details the timing when expenses will actually be paid. While labor payments could take place on a weekly basis, some of the other expenses could be paid on a monthly and/or even yearly basis. In medium to large projects where several hundred—if not thousands—of activities are budgeted and scheduled, such details may lead to substantial savings in financing costs.

While cash flows are essentially required for project financing, the timing of required resources is essential for the logistics of a project. Especially the distribution of budgeted critical resources over their task durations and placed in their time slots in the schedule provides invaluable information for PMs. Most planning software produces resource use reports, which is somehow the easier part of the process; preparing the required quantities and collecting the actual use at the job site may be more challenging.

In a similar way as its expense flow is generated, a project's resource flow can be established by replacing the costs with resource quantities for the activity cost components and spreading them along the project timeline as they will be used. The resource flow (i.e., schedule) would be most useful when it is prepared both on the activity basis and resource type basis. While the activity-based resource flow shows the resources needed for a specific activity, the resource type–based project resource flow provides what type of resources are needed for the entire project and when they will be needed. Such distribution of required resources reveals the peaks and valleys for resources along the project timeline, which may have to be adjusted by *resource levelling*.

As much as cash flow analysis is essential for project financing, major resource flows are key for timely procurement, recruitment, and logistics.

Preparing the Bid Price

If and when the budget preparing party needs a bid—or a selling—price, that can be easily calculated by adding a profit to the cost budget total. If there are any additional costs not yet included (i.e., bid and performance guarantees and other project-specific requirements), they need to be included in the bid price.

Some owners may require a partial or a full breakdown of the bid price. If that is the case, a good practice is to apply a profit coefficient to the costs of all WBS activities and regroup them for the required detail.

In preparing bid prices, a not so good practice is to *front load* some activities. Front loading is placing expected profit unevenly to activities that are scheduled to be performed earlier so that a better CF is generated during the early stages of a project. While that CF improvement alleviates the contractor's financing of the project at early stages, it increases the completion risk of later activities when the revenue for an activity may be less than its cost to the contractor. Owners not willing to make mobilization advance payments do contribute to this not so good practice frequently used by contractors.

Contractors are not project financiers. Owners need to understand that if their project has CF deficits (i.e., periods during which the project activities cannot generate the revenue to support operation/production) and if the contractors are not compensated for that deficit, they will have to pay for such either in monetary (i.e., inflated price) or in some quality/performance form.

In Summary

- A project cost budget is a plan with allocated resources.
- A project budget differs from an operational budget due to occurrence, duration covered, and since it requires unique resources.
- Project cost budgets include all known costs and allows for all known-unknowns in the form of a contingency or a provisional sum.
- Several budget formats serve their specific purposes in monitoring cost elements of cost components' line items and for projecting their costs at completion.
- Budget formats reporting past and current activity performances serve better for team leaders up to middle managers whereas formats projecting project completion may be of more use for middle to upper management.
- Budgets are periodically updated and revised when a predetermined threshold is exceeded.
- Cost budgets and time schedules are correlated. They are jointly used to generate project cash flow and resource schedules.
- An activity's costs may be incurred well before, during, and after the time slot in which that activity is carried out. That timing of incurring costs and the timing of receiving the revenue for that item needs to be studied in detail when preparing project cash flow.
- A project cost budget can easily be converted to a bid price with the inclusion of specific bidding expenses and profit.
- Front loading increases project completion risk; a mobilization advance against a guarantee alleviates contractors' finance needs at earlier stages of a project.

9 Controlling Cost

Introduction

Cost control is a major project task on its own. Depending on project size, a cost control department may accommodate several quantity surveyors, estimators, and staff for observing and reporting actual work achieved on various job sites.

Monitoring progress and taking action when necessary is the definition of controlling in management. Although widely and wrongly practiced, cost control is not reducing and/or minimizing cost. That may compromise the intended quality and/or value of project goals. Cost control monitors the performance indicators established in planning and provides corrective action and/or improvements by analyzing the root causes of variances. The control function of management needs to be proactive. The cost controlling process substantially differs from cost accounting, which is cost bookkeeping. By monitoring the cost elements, the cost controlling compares the actual quantities, unit costs, and production efficiencies to the estimated ones. Remedial action, improvements, and projections are then offered for monitored variances, providing early warnings of expected costs at project completion. When a cycle of the controlling process ends, it triggers the start of a new planning cycle. Planning— monitoring/feedback—taking corrective action/setting new goals defines the cycle of continuous improvement for a project. The continuous improvement does not end at project completion. It is carried forward to future projects with well documented *lessons learned* from completed projects.

This chapter covers details for selecting the activities, line items, and the cost elements to control, reporting frequencies and when to consider remedial action. Interpretations of monitored project status and root causes for cost variances in quantities, unit costs, and output performance, including various types of changes, are provided to enable readers to better understand their corrective action options are discussed. The importance of identifying the stakeholder accountable for the cost variance and how a stakeholder's remedial action may be governed by the delivery system and pricing option used for the project is emphasized. Remedial action decisions are categorized and explained. Corrective action disputes, dispute avoidance, and preferred settlement options are provided.

DOI: 10.1201/9781003172710-9

Cost Accounting Versus Cost Control

There are so many out there who confuse "cost accounting" and "cost control". This happens mostly because they relate costs to only dollar amounts and nothing else. Just like engaging lawyers to draft contracts, it is a common practice to engage accountants to control costs. Although accountants may well be capable of doing cost control, taking cost control as a by-product of accounting is not a good practice. *Accounting is the practice of keeping and presenting expense and revenue records for financial statements and for tax purposes in their pre-agreed formats.* The objective of accounting is keeping the officially required books for tax and other financial purposes. It needs to be noted that cost management is on its own a full-time job, especially for projects that require a full-time accounting.

Major Differences

By definition, "cost accounting" is a contradictory term on its own and especially when it is used in the context of "cost control". Accounting only registers "expenses" and not "costs". As explained earlier, when dealing with cost, managers should expect to see more than just a dollar amount. Cost data needs to be accompanied with a quantity and a dimension (i.e., unit) for a manager to make sense of the dollar amount it is associated with. More often than not, an expense is not of much interest to the manager from a "cost" perspective since it lacks the vital part of the information that the manager needs for proactively performing her/his management functions. Those management functions are controlling and feeding back to planning, organizing, and leading functions. In order to make a meaningful interpretation and decisions from the cost data of some specific work, at least the related scope, duration, and timing information must be analyzed together with the dollar amount attached to them.

Cost control is not an accounting practice. Accounting reports historic performance data, after they occur. Accounting is not designed to provide the much-needed alerts for cost overruns before they occur during the project. Controlling cost requires a proactive process and needs to use concurrent, feedback and feed-forward data.

Another common practice—difference—is using the general business account codes for monitoring costs of a project. Special attention needs to be given to adopting the existing account codes for a new project. Since every project is unique, from a cost standpoint the managers need to identify the cost priorities for a project. In the project WBS, some detail levels may match standard account codes used by accounting and may be suitable to monitor the cost for a specific case and/or activity. However, it is quite difficult to determine the cost significance of an activity and which cost component of what line item of that activity, without knowing the project's cost breakdown details. The subject account may include many component line items that may not really have cost significances and blending them all in one account may prevent the PM's observation of specific root causes of variance and determining what may be causing it.

As should be expected, offering a solution to an observed variance without clearly understanding its cause is problematic. Unclear causes delay PM's decisions for a remedy and makes the remedy harder and more complicated than it should be in the first place.

Frequency of reporting (i.e., monitoring) is one other issue that may be conflicting with accounting and cost management priorities. While issuing financial statements on a monthly basis may be totally sufficient for accounting purposes, daily monitoring and controlling may be absolutely necessary for some critical activities that may be completed way before the financial statements would be prepared. Again, it is worth noting that what may need to be monitored and adjusted or corrected may relate to quantities rather than the dollar amount. A project may require close monitoring of how much a specific material is used and how many labor hours and M/E hours are spent toward producing a certain product/service rather than focusing on the dollar amounts on the received invoices and paid payroll.

Accrued Expense Versus Committed Cost

Accounting registers and reports expenses which have to be either accrued or paid. Accountants need to receive an invoice and/or pay for a charge for them to register that expense in the books. When the organization establishes a commitment for some goods or services, the related cost is incurred by the organization. An organization (or the project manager as an agent) does not have to enter a written contract to make a commitment with a party to provide the goods and/or services. Even if there is a written contract, the deal may not require a payment made for a while and certainly will not require the supplier/provider to issue an invoice during that period. If there is no invoice and/or payment, there is no expense accrual related to this contract and hence there is no "cost" from an accounting perspective which is quite the opposite from the project manager's perspective. Managers have to monitor "costs" from the point of inception onward. Once "committed", the related cost is a fact that needs to be monitored and checked against not only what is included in the budget but also to what has been included in its elements; money and quantity-unit. As explained in detail in Chapter 8, such costs are reported as "Committed" costs in the project's cost budget.

In summary, the following should be considered before assigning cost management responsibility to accountants:

- Costs and expenses are conceptually different and accounting only deals with expenses.
- Accounting monitors expenses incurred and/or paid.
- Accountants do not estimate, plan, and monitor costs; they register and report expenses which by definition is not managing.
- Accounting feedback may not be in due time and specific enough to manage the costs effectively unless they undertake to divert from their routine practices to project specifics.

- Costs incur with or without an invoice and/or payment, which later turns into a payable debt. This "commitment" is not the usual accounting "payable"; it may or may not appear in an "invoice" form.
- Component quantities provide better future projection indicators for the remaining portion of an ongoing activity or for similar activities that will take place later in a project.
- To manage costs, material, labor, and/or M/E, quantities may need to be managed rather than the expensed dollar amounts for those components.

The focus here is managing the cost by planning and controlling the cost rather than "accounting" for it. The only "control" accounting can offer for an expense is mostly limited to the processing speed of registering that expense, cutting checks, and preparing the financial statements. While that is a well-respected service, managers need a little more than that to properly keep their project budget, schedule, and cash flow on track.

Step-by-Step Cost Controlling Process

The controlling function of management can be wrapped in two main subsections. Monitoring the planned activities is the first section and taking corrective action to remedy significant deviations from the set targets is the second. Controlling is initiated with planning where the activity goals are established together with their performance criteria. A good plan always establishes attainable and controllable targets at its inception. If a target is set without the performance criteria that allows objective evaluation along activity progress, that activity can be monitored but cannot be controlled. If a manager cannot control a plan, that manager cannot manage that project; he/she can only monitor it. The evaluation for quantitative constraints such as scope, cost, and time may be considered easier since they are based on numbers. The evaluation of qualitative constraints (i.e., customer satisfaction, team building, brand development, environmental impact, and similar others) on the other hand, may require more elaborate evaluation processes such as interviews and surveys which are beyond the primary scope of this book.

Figure 9–1 shows the schematics and summarizes a typical project cost controlling process.

While all project activities need to be included in cost planning, controlling may not be needed for all of them. All activities needed to get the job done may not have the same significance from an overall project cost goals perspective. Analogically, as much as we would like to have "salt" included in the right prescribed amount in a banana bread recipe, overusing it will only create a "quality" issue and not a major cost overrun as its cost analysis will show. While controlling all activities may sound like a good idea, it may not be practically possible and it may not serve the purpose of making it easier to manage project costs.

Establishing the significance—and prioritization—of the budget line items to be monitored will start by an order of magnitude ranking. That may be done by simply giving the top priority to the highest dollar amount item on the budget. When

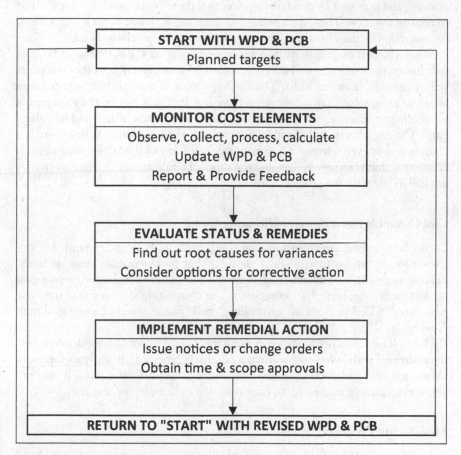

Figure 9–1 Cost controlling roadmap.

priorities are established for the budget line items, the criteria as to when to take corrective action for this line item—just like the case is for the entire budget—should also be determined. The most common thresholds to revise a budget line item would be 10~15% and/or a predetermined dollar amount as the variation limit. The purpose of cost control is to take proactive action to prevent reaching and/or exceeding those thresholds. Setting up the priority activities and the thresholds does not mean that priorities and the criteria to take corrective action cannot be modified at the judgment of the project manager while the project is in progress.

When budget line items are prioritized with their order of cost magnitude, a controllability study needs to take place before finalizing the list of line items to be monitored for cost control. The controllability of a cost line item is the PM's ability/authority in controlling cost elements of any line item. If the quantity and/or unit cost of a material component line item is fixed (i.e., quantity by specification/

drawing and unit cost by purchase agreement), there is not much left for the PM to control that cost. There is no point in monitoring such line items if they cannot be controlled by the PM, although their total cost may be relatively high.

When the cost of an activity deviates from its original value, that would trigger a deviation from its estimated duration and/or its requirement for other resources (i.e., materials, labor, or M/E). The consequences of a corrective action taken could be acceptable from a cost perspective but that may not be the case from a scheduling perspective since that activity may change the schedule and its critical path. The scope, cost, and duration of a project are interrelated. When there is a change and/or even a minor adjustment made for any of the triple constraints, it is essential that both cost budget and the time schedule are verified to ensure they are still in line with the project's targets.

Cost Control Is Not a Process of Reducing Cost

Controlling is a management function that enable PMs to understand the root causes for deviations from original plans so that proper management decision-making steps can be implemented to assure the planned or better project goal achievements. Applying that definition to cost controlling, it should read the same with "cost" added in front of "controlling" and "plans" should be replaced with "cost budgets."

Controlling cost is not a process of reducing costs. Project cost reduction is better addressed with value engineering, a wholistic process that analyzes function, value, and cost and offers optimizations where possible. Cost control is also not about practicing *position power* to limit resource availability for activities.

Monitoring Progress

Progress monitoring is the first leg of controlling process. It involves tracking and reporting accomplished activity progresses for specific periods. Depending on the project size, duration, and organization levels, such monitoring may require multiple levels of tracking and reporting. A key decision for monitoring is the period of reporting, which may differ for reports prepared for upper management and for different project stakeholders. While the first line managers may need to monitor progress on shorter periods and in full detail for their scope sections, less frequent and less detailed but more comprehensive reports will be more useful for middle and upper managers who are in a position to see the entirety of the project.

All cost elements need to be monitored in a cost control process. Quantities as well as unit costs obtained from purchase orders/agreements (i.e., committed costs), procurement (i.e., invoices) and output efficiencies (i.e., performance/progress achievements) need to be monitored. This requires calculating the quantity of actual work done and sorting out resources consumed and payments made for specific tasks as they are defined in the WPD. Since materials are not ordered on a task basis, a received invoice may need to be distributed to several activities so their respective costs are properly allocated. Similarly, labor trades—and/or

M/E—can work simultaneously on more than one task. Sorting and allocating the right cost data for such tasks requires good knowledge and attention to the WPD and what is targeted to be monitored.

Monitoring includes updating cost budgets by using collected, calculated, and allocated data to the tasks to which they belong. The updated cost budgets and any expected duration change projections are to be distributed to the organization as well as other stakeholders in the appropriate format and detail.

What to Monitor

In a good size project where the WBS may contain several hundreds if not thousands of tasks, monitoring all those tasks might not be practically viable. In addition, that number of activities can easily fold several times when specific line items and/or components of line items and/or the cost elements of those line items are considered to be monitored. For effectiveness and efficiency of the process, a prioritized critical list which will best serve the specifics of the project needs to be prepared. All activities that may potentially impact the cost outcome of a project make it to an initial cost monitoring list. That list is simply a rank order list based on the cost magnitude of activities or parts of activities such as the cost line items of activities. The initial cost monitoring priority list is expected to reduce the number of tasks to what is realistically manageable by project staff. Selecting to monitor activities which do not have cost significance will only result in inefficient use of project resources.

Due to staff shortness, if the initial short list has to be so short that in the opinion of the PM it is not including all critical activities, additional staff needs to be required and appointed so that all cost critical items make it to the list of items to be monitored. The additional staff may include project site aids who would observe, collect data, and assist team leaders in reporting that data for further processing at higher management levels.

The process of selecting the cost line items to monitor starts with choosing the ones with relatively higher costs but there is more to consider. Determining what is *controllable* is an equally valuable assessment when selecting what needs to be monitored. Once an initial cost monitoring list is prepared by eliminating magnitude-wise less critical activities, a closer look into the cost details of the items on that list will reveal what needs to be monitored. What needs to be monitored is determined by scrutinizing what can be *controlled at the time*. In theory what can be controlled are the cost elements, quantities, and unit costs for materials and production output rates for labor and M/E components of these major cost activities. Although theoretically all cost elements should be controllable, that may not be the case for a specific task. Unit costs of materials and hourly wages that are determined by sales and employment agreements are examples of cost elements which are practically not controllable at the job site. For such cases, material use and output rates needs to be focused on.

Major activities of a project are not necessarily *critical* activities to be controlled from a cost perspective. In addition, all cost component line items of major cost activities may not have the same cost impact (i.e., not equally critical) on the project. Table 9–1 depicts selected priority cost element line items for the floor slab, the estimate of which was given in Chapter 6, Table 6–6. The PM makes a judgment call to determine which line items to monitor. These priorities may change anytime the PM deems necessary during the course of the project.

Table 9–1 lists the line items of an assumed to be a cost-critical activity for a project and shows the line items selected to be monitored for controlling their costs. A closer look into the example presented in Table 9–1 reveals some facts that may look contradictory at first glance. The selection of the major cost activities in Table 9–1 are not done by simply sorting the overall cost total of activities. Considering the magnitude of costs, there are two material component elements and two labor component elements that appear to have relatively high costs in Table 9–1. The totals for both of the material component line items, namely the "ready mix concrete including transportation and pump" and "rebar delivered, cut, bent, and tagged" are higher than the estimated labor costs of carpenters and laborers. However, both materials are not selected as cost line items to be controlled. In this example the selection of what activity to control is done on the basis of how effective a control can be performed on these activities. The suppliers of these materials, will produce and deliver exactly what is ordered. Controlling the quantity delivered and the unit price charged should not be a cost control practice but should be a procurement delivery control practice. The unit costs for these two material line items include all production and delivery costs, leaving only the ordered quantity and the unit cost to be the two elements the receiving party can control. One would assume that the unit costs for such activities are negotiated and committed for the entire project. So these unit costs, negotiated and contracted, cannot be controlled specifically for this activity. The quantities on the other hand are fixed and provided to the supplier by the project organization and are calculated by using the relevant design drawings. The only controllable variable for the material quantities of these two line items would be calculation and/or application errors. For the concrete, an application error stemming from the measuring of the forms at the time when they are erected in place is possible. If all measurements were off by ¼", the resulting concrete volume error for this particular case is calculated to be in the order of 20 cubic yards. The selection of cost-significant activities to control would be totally different if this project produced its own concrete and cut and bent its own reinforcing steel on site.

The costs for the two labor line items, on the other hand, can vary based on a number of factors. Among many others, selected construction method, team formation, training, efficiency, equipment compatibility/capacity, and leadership skills can be listed as examples of factors that can impact the resulting cost for these line items. Obviously, such factors are not of concern for the concrete and rebar material line items.

Another approach to selecting activities to monitor can be made through analyzing Cost at Completion (C@C or CAC) projections in the budget. By analyzing the cost data reported on previous and current periods, a PM will have a good understanding of how the costs of certain activities are trending. An unfavorable

Table 9–1 Prioritized cost line items for reinforced concrete floor slab

Component	Description	qty.	unit	unit cost ($)	total cost ($)	Monitoring priority
MATERIALS	**Ready mix concrete** incl. transprt. and pump	273	yd³	175	47,775	(*)
	Forms				8,070	
	Metal props for deck (100 uses expected, 1/100 cost charged)	700	each	2	1400	
	4x4" × 8' beams (10 uses)	700	each	2	1400	
	3/4" plywood (5 uses)	480	each	8	3840	
	2x4" × 8' wood (10 uses)	300	each	1	360	
	Wood planks 2 × 8" × 10' (10 uses)	100	pieces	1	50	
	Pegs 1 × 2 × 12"	1	box	20	20	
	Form oil	100	gallons	6	600	
	Nails, screws, tie rods, braces, and other consumables	1	l.s.	400	400	
	Reinforcement				57,225	(*)
	Rebar delivered cut, bent, and tagged	18	tons	3,000	54,600	
	Rebar wire	500	lbs.	1	625	
	Spacers	400	boxes	5	2,000	
	Miscellaneous				4,600	
	MEP	1	l.s.	2,500	2,500	
	Anchors	1	l.s.	1,500	1,500	
	Plastic moisture barrier	14,000	sq.ft.	0.100	1,400	
	Burlap for water curing	14,000	sq.ft.	0.200	2,800	
	Water for curing	100,000	gallons	0.004	400	
	Total Materials				117,670	

(Continued)

Table 9.1 (Continued)

Component	Description	qty.	unit	unit cost ($)	total cost ($)	Monitoring priority
	General Foremen	110	hours	60	6,600	(**)
	Carpenter	766	hours	40	30,640	
	Iron worker	100	hours	40	4,000	(**)
LABOR	Laborer	1,732	hours	25	43,300	
	Electrician	24	hours	45	1,080	
	Plumber	32	hours	45	1,440	
	Labor Total				**87,060**	
MEOD	The site installations, storage/staging areas, vertical and horizontal transportation equipment, and similar other resources are used but their cost is not included. The general site operation cost will be calculated and added as a general overhead. Only specific handtools, lighting, power extensions, and other similar tools/equipment provided are included here.	1	l.s.	1,000	1,000	
	MEOD Total				**1,000**	
O/H	Only task-specific items such as specific gear, cleaning and repair supplies are included here.	1	l.s.	750	750	
	Total Overheads				**750**	
	TOTAL				**$ 206,480**	

(*): Not prioritized as explained in the text.
(**): Prioritized tasks for cost controlling purposes.

trend is a red flag. Based on that alert, a PM may decide to include those on the activities to be monitored list as well. Although it is an alert, an unfavorable cost trend in any period on its own is not conclusive for a budget overrun if the C@C is still within the budget.

Monitoring Frequency

With proper selection of what needs to be monitored, the focus shifts to monitoring frequency, which depends on what is being monitored and at what WBS level. The most helpful tool for determining monitoring frequencies is the project WBS. It reveals the size, organization levels, and the complexity of a project. The WBS also indicates the reporting structure of the project. While the more frequent reporting may be required at lower management levels, that frequency will reduce as the reporting levels go higher. Although daily monitoring is very common, monitoring frequency may go as high as hourly reports at the first line management (i.e., team leadership) level. Weekly, monthly, and/or quarterly monitoring are more common at middle, upper, and top management levels. Similarly, reporting the monitored progress to different project stakeholders, depending on their role and position in the project, may differ substantially.

Progress monitoring frequency also varies according to project size, complexity, duration, cost, and on the levels of organization and number of stakeholders. A *big* project will take longer, will cost more, will require a bigger—more layered—organization, and may have more stakeholder involvement. The size, whether the project is big or small, is a relative concept. What could be a big project for one organization could be considered a very small project by another. One other factor, project complexity, is also an organization-specific concept. What could be considered as a high complexity project for one organization could be a piece of cake for another. So, how can one decide on the monitoring frequency to use? The answer is, there cannot be one cookie cutter solution for each and every project. Each project being unique dictates that each project will have its own, unique monitoring frequencies.

Owners usually monitor their construction projects on a monthly basis. The monitoring frequency increases for the contractors and as the monitoring is carried out for smaller sections of work. Contractors' field staff monitor tasks on a daily basis.

How to Monitor

Monitoring involves in-depth understanding of the plan details. Since the subject plan is the cost budget, all managers, including the team leaders, need to be provided with full details and how they are expected to track (i.e., observe) and report the physical progress as well as the resources used to achieve that progress. In return, managers will review their assigned task(s) in full detail with their team members and come up with their plan for accomplishing them. At this point, everyone involved needs to know clearly what needs to be delivered, using what resources, and for how long as well as how frequently they are expected to report the progress on physical quantities achieved and resources (i.e., materials, labor trade-hours, and M/E hours) used.

Monitoring is best reflected on a pre-agreed reporting format which will be the least time- consuming to prepare and clearly communicate progress with the organization and other stakeholders.

Reporting

Monitoring involves observation of what is happening on the project and conveying that to others involved in that project. Conveying the message, the communication through reporting, needs to be *noise free* and *effective*. *Noise* is anything that interferes with the intended message of communication. A communication is *effective* when the intended message is clearly understood by the receiver. The key factor for noise-free and effective communication is standardization of the message. This is best obtained by periodical, standardized reporting by using predetermined formats.

Reporting is done by using the organization's formal communication channels. The written medium is always the most preferred, although for some rare cases (i.e., emergencies) and temporarily, oral communication can also be allowed. The preferred written medium is digital where applicable. Depending on internet/satellite service availability and reliability, hardcopies may also be required.

Managing a project mandates issuing and receiving several types of reports with varying frequencies and content. Reports are prepared at several management levels for higher management. The frequency is higher with specific task details at lower management level reports. The frequency gets lower and the coverage widens as the receiving management level gets higher. The length of a report is an irrelevant statistic. Longer reports do not necessarily express the messages better. The effective conciseness is a much sought after quality when it comes to reporting. Reports prepared for other project stakeholders are prepared for the top management levels of those stakeholders.

Timing of reporting is as important as its accuracy. A delayed report is as unacceptable as an inaccurate report. Some managers may need to be educated that delivering the product (i.e., scope) is not their only target. They have to deliver in due time and within its targeted cost. If a manager does not have time to report what he/she has worked on, that manager should be considered to be having issues managing their assignments.

DAILY REPORTS

Daily reports (and shift reports when applicable) provide information on what has been accomplished during the workday. The first line managers prepare daily reports. One or two level higher managers might also produce daily reports depending on the project.

Appendix C.1 provides a sample daily report form.

Starting with the date, weather, high and low temperatures, number of workers, work done, materials and equipment, deliveries received, orders placed, safety issues, accidents/incidents, any observation/warning/recommendation for improvements/shortcomings, and other similar issues that relate to that day's operation are included in a daily report. Efficiently formatted, most daily reports take a couple of pages to present all that information. Additional pages are used to provide specifics for adverse situations (i.e., accidents, fires, etc.) and for specific feedback when necessary.

Daily reports should not be difficult to prepare. That is easily achieved with training the teams on what needs to be reported. The receiving manager should always be open to improvements that will ease the preparation process. If any part of the report is practically too difficult to perform on a daily basis, it is a good practice to have a reminder to indicate when such reporting will actually take place. An example of such a case could be made on some materials. For practical reasons it may not be easy to report actual quantities of *all* materials used on a daily basis. A reminder to check critical material stock on another frequency (i.e., weekly, biweekly, monthly, etc.) to make sure that replenishments are ordered in due time could suffice on a daily basis.

Depending on the priorities of the project, daily reports may require more detailed information on certain cost components of activities. One common application is to use daily reports to capture labor hours of several disciplines for the WBS tasks those hours relate to. Such collected data is essential in calculating labor output rates and efficiencies that will impact the final cost of the related activities. One other use of such data is when similar work is carried out by a number of separate teams. The actual labor hour per unit production of one team compared to that of others is a good indicator for team–leader efficiency and performance. Such data driven from the daily reports become the best estimating reference for future similar work.

PROJECT PROGRESS REPORTS

In any organization where there are several managerial reporting levels, the PMs of those organizations produce progress reports for their superiors and for other stakeholders as may be contractually required or necessary. Most commonly, project progress reports are issued on a monthly basis. These reports are officially documented communications between managerial levels and between stakeholders. They present what has been accomplished with what resources and make projections based on monitored data including daily reports from team leaders as well as data gathered from other departments (i.e., accounting, design, procurement, etc.) and convey PMs' messages about the project. Monthly reports are a means for the PM to communicate his/her view of the current and future needs and concerns of the project they are in charge of managing.

An executive summary, project fact sheet, cost and resource budgets, revenues, cash flow, schedule, physical progress (i.e., critical quantities achieved, labor and M/E hours used), performance indicators, conclusion, and remarks are usual sections of a project progress report.

A sample form for a monthly progress report is provided in Appendix C.2.

Processing Progress, Invoices, and Payroll

The cost controller receives all cost-related data from wherever some cost is committed, incurred, and/or accrued. The received cost data includes the reports from the job site and all invoices and expense reports from accounting and payroll. These reports need to be processed so that respective costs are allocated appropriately to their related tasks. The processing of cost data also includes the comparison of the received cost data with what is in the cost budget. This is done by separately comparing the two cost elements, the dollar amounts and the related quantities of planned versus actuals. While what is planned is—should be—available in the WPD, PCB, and the cost budget, the values for actuals (i.e., the achieved values) can be obtained by *Sorting and Distributing* the received cost data and by *Calculating/Verifying Actual Quantities* of the prioritized tasks.

SORTING AND DISTRIBUTING ACCOUNTING COST DATA

The reports the cost controller receives from the accounting department includes *all* invoices, expense reports, and any means accounting receives cost data. The data received needs to be processed so that respective costs are allocated for the intended tasks. To make the right cost allocation, the reported data may require sorting and/or further quantity calculations. The most common examples of this process are material cost allocations and labor output calculations. Since one material may be used in more than one activity and since the suppliers issue their invoices based on total quantity delivered, a received invoice total needs to be distributed to all tasks that use that invoiced material. Sorting involves the tasks that have used the invoiced material whereas the distribution of the cost involves the quantities of that material each task has used.

The labor trade hours and M/E hours also need to be sorted and distributed before they can be associated with a specific task and quantity of work achieved for that task.

The calculated work achieved and resource hours used form the basis for verifying output rates of teams and M/E.

CALCULATING/VERIFYING ACTUAL QUANTITIES

The actual quantity of work achieved is expected to be reported by the team leaders performing the tasks. The reporting managers will do so by measuring and calculating the quantities achieved. Receiving such quantities, the cost control office will have to verify them by comparing them to the BOQ, WPD, and/or PCB quantities as well as with the billed quantities.

Interpreting Monitored Performance

The monitored progress achieved at a certain point in a project timeline will place the project status in one of the nine cost versus time zones shown in Figure 9–2.

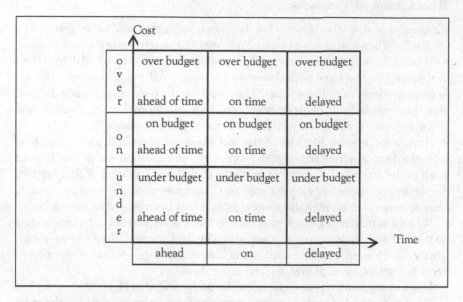

Figure 9–2 Project progress status in nine cost–time zones.

Every zone shown in Figure 9–2 conveys a message from scope, cost, and time perspectives. Monitored project status indicating that it is "under budget and ahead of schedule" is a very good place to be for the PM. While the teams and the PM need to be praised for that performance, other underlying factors also need to be investigated. The question that needs an answer is whether this is a result of overestimated scope, cost, and duration. If not, the conclusion is admirable. On the other extreme, if the project is over budget and behind schedule, underestimated quantities, cost, and durations become the first things to look into. For other zones where cost and time status indicates opposing directions, again the underlying reasons are of utmost interest for the PM. Why does one constraint do fine while the other does the opposite is a concern that the PM is in a position to address.

Some common underlying reasons for such project status are:

- Ahead of schedule and over budget: Use of excessive and unnecessary overtime
- Behind schedule and under budget: Inexpensive resources producing less than desired output
- Behind schedule and over budget: Excessive overtime and/or expensive resources producing less than desired output

There may also be a number of other reasons for a project to deviate from its targeted cost and duration. The possible reasons for those deviations are presented next.

Root Causes of Variances

Variance is a deviation from what has been estimated and/or targeted (i.e., planned). When a significant variance is monitored, determining the root cause(s) for that deviation becomes essential for making the right decision(s) for remedial action(s). The most critical step in the managerial problem-solving process is properly identifying the problem. The plurals in the previous sentence indicate that there may be more than one root cause, requiring more than one decision and involving more than one corrective action for a specific variance.

A variation observed in one of the cost elements of an activity is an indication of a deviation from the original scope, cost, and duration which may present itself in the estimated quantity or unit cost of any cost line item of that activity. Depending on the activity significance and the significance of the variation itself, that activity variation may also cause a project cost overrun and/or time delay.

When it is monitored, the reason behind that variance needs to be determined so that the appropriate corrective action can be implemented. Root causes of variances may be attributable to any stakeholder, introducing another dimension to remedial action decisions and implementing them.

Major cost variation root cause examples are provided in Figure 9–3.

Details of indicated root causes and examples for them are provided in the order of their appearance in Figure 9–3.

Observed variation in:	Probable Cause
Quantity	Take-off error Scope creep Excessive waste Damages Changes[1] Unforeseen (adverse) conditions
Unit Costs	Estimating error Inflation/market conditions Changes[1] Unforeseen (adverse) conditions
Labor and M/E Output (performance)	Estimating error Utilized method compatibility Workforce skill/training Teamwork issues Leadership Organization Changes[1] Unforeseen (adverse) conditions

[1] See Figure 9–4 for details

Figure 9–3 Root causes of variances for cost elements and output estimations.

Common Reasons for Quantity Variance

Quantities relate to measuring the scope of work. The reason for a quantity variance may be as simple as a human mistake or a major unforeseen condition. This section presents some common probable causes for scope/quantity variances. The quantity variance may also be encountered due to scope changes, which is explained in the changes section that follows.

Take-Off Errors

Estimating errors include simple calculation mistakes, failing to include a section of scope, duplicate inclusions of scope sections, and over- and/or underestimations. When he or she becomes aware, the PM immediately reports such findings and requests a budget revision authorization and a contract price adjustment where applicable. Since the duration of such activities will also be impacted, a schedule revision may also be required. However, if there are no estimating errors based on the documents used but a variance exists between what was estimated and what is actually accomplished (i.e., delivered), the correctness of the documents used for estimating the cost and the documents used to construct the deliverable becomes questionable. More than often, the designers tweak their designs without paying attention that even a small tweak may cause an unexpectedly higher cost. Another reason for a cost deviation could be a conflict between documents such as a drawing showing a size that is prohibited by the specifications. Such cases may require processing change orders to compensate for the additional cost and extending the project completion duration accordingly. Additional work compensation and time extension claims become problematic unless they are dealt with at the very early stages of the change. Since working out the entire cost impact may take a while, notices informing the stakeholders of such changes are extremely important and, in most cases, legally required within articulated time limitations.

Scope Creep

The incremental increase in the scope starting with one minor change leading to other scope changes is one definition of scope creep. Poorly defined, unrealistically specified scope is the most common cause of scope creep. It is the process of completing the scope definition as the work progresses. Scope *creep* implies that the corresponding changes for cost and duration are not accounted for. Otherwise, it would be a *change order*, which is expected to address and compensate for relevant cost and time.

Excessive Waste

Excessive waste may be a direct result of the method used and the workmanship skill level. The material quantity requirements (and labor and M/E outputs) are to be carefully studied since they may have significant differences, especially when there are unfavorable quantity and resource compatibility issues. A large paint sprayer may require a quarter of a gallon of paint to be prepared to spray. If this equipment is used to spray paint over an area which could be covered by, say, half a gallon of paint, that operation will result in 50% of paint wastage only due to poor selection of method and M/E to implement it.

Lack of training and unqualified labor are also reasons that lead to excessive wastage. Other causes of wastage include poor storage/staging facilities, vertical and horizontal site transportation means, and other adverse working conditions on site.

Damages

Damage to work should be expected due to poor protection and maintenance of work as well as accidents at the workplace. Damages not only increase the quantity of work but also diminishes end product quality. Special protection precautions and measures are to be considered for tasks when they are expected to be completed while other activities continue in the vicinity or different labor trades have to work on the same activity, one starting after the other finishes its part.

Adverse Conditions

This is a category that may be construed in more than one way. It contributes to variations of quantities, unit costs, and output for labor and M/E (i.e., performance). It is a change to what has been assumed/targeted. The consequences of such conditions (i.e., this type of change) are defined by the delivery method/contract pricing combination used that assigns specific obligations/responsibilities to stakeholders.

Adverse conditions are those which have not been specified and/or are conditions that a prudent estimator could not have foreseen at the time when the estimate was prepared. Adverse conditions examples are extreme hot/cold temperatures, high winds, floods, and other extreme meteorological events. Such conditions may impact quantities of activities, direct unit costs (i.e., due to transportation, delivery difficulties, and production inefficiencies), and indirect unit costs (i.e., due to increased supervision and additional precautions required). Adverse conditions are not as severe as force majeure conditions; they make it harder to carry out the work whereas force majeure conditions shut down the operation (i.e., make it impossible to work safely).

Changes

> "Nothing is permanent except change."
>
> Heraclitus, 575–435BCE

Changes are inevitable for any project. Changes happen due to an array of reasons. When they happen, the worst thing do is to ignore the change. When changes are managed properly, their impacts are minimized and those impacted projects are given a chance to get back on track. When ignored, even a minor change may escalate to cause a project's total failure. Change Orders impact direct and indirect costs and activity and project durations. They have further and differing implications on the stakeholders depending on the delivery method and contract pricing used.

Some common types of changes are shown in Figure 9–4, details of which are provided below.

SCOPE CHANGES

Scope changes are additions and/or omissions to the original scope. Additional work could be an *increased* amount of work for an existing activity or a totally *new work* may be added to the original project scope. Likewise, a portion of an activity or an entire section of work may be omitted. Scope changes may also originate from alterations or even conceptual changes to design.

DESIGN CHANGES

Design changes become inevitable when the current design has deficiencies or it cannot be accommodated and/or the current design no longer fulfills the business intent of the project. Constructability reviews and value engineering may result in design changes as well. Design changes may have significant cost and duration implications if they take place after orders are placed and after related production/construction works commence.

CHANGES IN REGULATIONS

Regulation changes can happen anytime and anyplace. Usually such new regulations grandfather projects that have already been permitted. However, changes in tax legislation, import/export regulations, special security, and public health safety

Changes due to:	Typical Examples
Scope	additions/omissions of work
Design	development, concept change, coordination, conflicts, specifications
Regulations	codes, permits, licensing, taxes, labor related regulations
Differing site/workplace conditions	access, utilities, safety/security, unforeseen conditions
Failure of Owner/other stakeholders	approvals, deliverables, payments, site handover
Force majeure	"acts of God"; adverse conditions beyond any stakeholder's control

Figure 9–4 Change categories and typical examples.

regulations may have a direct and/or indirect impact on existing project quantities, unit prices, and indirect costs.

DIFFERING/UNFORESEEN SITE CONDITIONS

If the scope and unit costs of the original site conditions assumptions are not validated, they need to be changed as the site conditions dictate. Encountering a higher than specified water table at an excavation site and the lack of an access road to the job site are common examples of such conditions. From the utilized delivery method/contract pricing standpoint, differing and/or unforeseen site conditions impose stakeholder responsibilities similar to what is the case for adverse conditions.

FAILURE OF A STAKEHOLDER

Failure of a stakeholder will prevent other stakeholders from fulfilling their contractual obligations by preventing them from conducting their businesses and causing them to incur additional costs for delays and extended project duration. Owners not being able to provide information that is essential for designers, being unable to obtain permits, not paying and/or delaying payments, and delaying approvals are some issues that cause delays and consequent change orders. Designer failures are usually delayed issuance of drawings and specifications and producing defective and/or conflicting design documents.

Force Majeure

This is a general term used for defining unfavorable situations that are not attributable to any stakeholder. Natural disasters, riots, labor strikes, and wars are examples of force majeure where due to no fault of any party a project is stopped, harmed, and/or delayed. The conditions included in force majeure are not project specific; they impact the entire project location.

Unit Cost Variance

Variation to estimated unit costs may occur due to several causes as listed in Figure 9–3. While some of these causes (i.e., errors and adverse conditions) are obvious, some others require special attention to be revealed. In addition to the ones included next, some other causes like the technique used, organizing teams and selecting team members, training, diversity, groupthink, and leadership styles all have their own contribution to output inefficiencies and hence to the units costs of tasks.

Errors, Inflation, Changes, Market, and Adverse Conditions

These are all factors that may be well hidden at the time of cost planning and partially during the execution of the work. They all may directly impact relevant unit costs. Inflation and market conditions may not even be observed in the day-to-day project monitoring operations. Periodic checking with the sources used during estimating and planning is the best proactive way of finding out the trends for inflation and market conditions. Revealing errors and changes mostly depends on the frequency and the accuracy of reporting. Correlating adverse conditions to increasing unit costs is more difficult, but again, good monitoring and record keeping will provide the needed statistics if a case were to be made.

Unit Cost Variance and Quantity Implications

A significant quantity variation may have an impact on the unit cost of resources required to conduct the related activities. While increased quantities are generally expected to lower the unit costs, the opposite is also valid in some specific cases. Suppliers usually lower their prices at the opportunity to clear their stocks. At the time of procurement, the supplier might give a good discount for a large order and/ or if the order quantity clears their stocks. At a later stage, if the supplier does not have enough stock for the required additional quantity, producing an additional batch may become necessary. The new batch production may involve updated prices and production and delivery durations, all of which may incur additional costs. In this case, increased quantities may result in increased unit costs and delays.

Inflation is another concern which may lead to increased unit costs during the project.

Extended labor and M/E hours may require additional expenses such as overtime payments and additional M/E that are generally named as *acceleration* costs. When the variation to the scope is bigger than what can be managed by overtime work, new teams and M/E will be required. Their direct and indirect (i.e., recruitment, transportation, etc.) cost aside, having to use such additional resources will most probably impact the activity duration. An extended—or expedited—task duration may shift the project's critical path, which may impact other activity costs and durations. In addition, idle time incurs fixed (i.e., owning costs) or variable (i.e., rent) costs during periods where M/E is not operational on the job site.

Project delays increase O/H costs since most of these costs are directly related to overall project duration. O/H continues to incur during all times whether the project is active or not.

As discussed earlier, adverse and differing site conditions also contribute to variations of unit costs.

Labor and M/E Performance Variances

Chapter 6 discusses the labor and M/E component output in detail. Figure 9–3 provides the most common causes for variance in these two component efficiencies. By definition, a project is a unique endeavor and unique situations cannot be fully covered by standard checklists. This is especially true when the subject is labor efficiency, which may be directly related to human behavior interactions of team members and leaders. Team formation techniques and matching leadership styles are proven to substantially increase output rates, but unfortunately, they are rarely used in the construction industry. Other than those that are behavior and personality related, the following are the most common labor and M/E performance deficit causes for construction projects.

*   **Estimating errors** for performance deficiencies include wrong and/or overoptimistic output rates.
*   **Utilized method compatibility** relates to the selected construction method not being the best option for the job location, site conditions, season, and/or weather conditions and similar other project specifics.
*   **Workforce skill and M/E capacity compatibility** impacts the output when labor skills/training and selected M/E to carry out the job are not compatible with each other. In addition, the selected method may also not be compatible with one or both of these resources or vice versa.
*   **Teamwork issues** include selecting complementing team members, forming, and norming of teams. While an individual team could be successful on its own, working in synchronization with other team members and/or teams may be an issue.
*   **Organization–workflow issues** negatively affect the output if the organization does not provide an efficient work flow for all involved resources. The workflow efficiency also depends on the accuracy of timing (i.e., how long it takes) for each resource to perform their part of the task.
*   **Leadership** impact on team output has been the subject of extensive research. It can on its own be a factor limiting the expected output. This is a subject that has not been studied in the construction industry as well as it has been in other industries. Considering the purpose of this book, only mentioning the fact that a mismatch between leadership style and teams formed may have significant negative impact on the output is included here.
*   **Changes** do have a wide spectrum of impact on output. The interruption of existing workflow, remobilizing, necessity for different method and resources, and down time awaiting resources are all among many other factors affecting output.
*   **Adverse and differing site conditions** are those conditions that are either unaccounted for extraordinary conditions encountered or wrongly specified conditions that do not match the reality on site. Unusually extreme meteorologic conditions, lack of access to the work site, and encountering different than specified soil conditions are typical examples of conditions which are either wrongly specified or could not have been foreseen by experienced designers and builders. Dealing with the consequences of such conditions will require additional and more costly resources due to change of method, M/E type, and capacity, additional and different labor trades, and specialty contractors. These conditions are project-specific conditions which are different than conditions included in *force majeure*. Force majeure is not project-specific; it is a general condition that impacts the project as well as the community of the project site.

Remedial Action Considerations

Remedial action considerations for cost deviations can generally be positioned to address the root causes in categories as outlined in Figure 9–3. Any remedial action for any root cause category can further be broken down to:

- **What** needs to be done,
- **How** will it be implemented, and
- **Who** will compensate for it.

Finding simple answers to these questions for quantity, unit cost, and labor and M/E output categories are not always easy. While the answer to *what needs to be done* can be as simple as increasing resources, answering the following *how* and *who compensates* questions may get extremely complicated.

Finding the right remedial action and implementing that action will need to be custom designed for specific cases of every project. One solution that works fine for one project may not produce the same results in another due to what may seem to be a subtle difference for one stakeholder but not for other(s). Unless the ramifications of a remedial action is self-contained and does not affect others' project targets, the considered action needs to be discussed and agreed to by all stakeholders.

Cost overruns cause both financial burden and delay in project completion. Hence the remedial actions most probably will involve cost and time compensation which may present substantially differing ramifications for stakeholders. An owner may accommodate compensating for time but compensating the cost may reverse feasibility of the investment. Or the same owner may have the funds to compensate for the cost but cannot tolerate any completion delays. Similar scenarios can also be developed for the contractor whose priority in general is the cost rather than the duration. All in all, neither of the stakeholders would like to suffer additional cost and time. Determining the party to compensate for the deviations depends on a number of project specifics and is an extremely sophisticated process. It involves delivery system, contract pricing, and the priorities and professionalism and organizational cultures of stakeholders, all in the context of project specifics.

Variance Accountability and Remedial Action

Identifying the stakeholder whom the variance is attributable to is a key assessment toward managing the effects of variances. Depending on the *delivery system* implemented, project stakeholders have their own priorities and obligations with regard to project scope–cost–time limitations. Of the three fundamental stakeholders, the owner bears the highest risk when such requirements cannot be complied with. Even if the contractors and designers can be penalized for their underperformance, the owner suffers the most when a project cannot meet its original goals. The penalties, even in the form of *liquidated damages* cannot even come close to compensating the owner's losses when a project is delayed and/or has to be abandoned. For the owner, a project is a vehicle to advance business goals. Completion of a project is only the start of generating a stream of benefits for the owner which is expected to last far longer than the project duration itself. The penalties applied to other parties can merely compensate for the losses an owner incurs when a project is derailed from its targets.

Intended or unintended, the owner may initiate changes that can significantly impact project cost and duration. It is therefore incumbent on the PMs of all stakeholders to monitor the progress and practice *feedforward control* so that performance deficiencies can be prevented before they can be detrimental. Especially when a project is executed in house (i.e., totally with the owner's own resources and not in contract with other parties), identifying who caused the change and pointing out misjudgment and errors become even harder.

It is always difficult to pinpoint the responsible party and/or specific action that starts a snowball effect which can derail project targets. Since activities are interdependent from all aspects of scope, cost, and duration, a direct or indirect change to one activity may start that snowball rolling. The ripple effect of that variance will likely have consequential impact on other activities. The best practice is to make everyone aware of what the change is, what trend it represents, what consequences are to be expected, and who will be responsible for the consequences as early as the first monitoring frequency the variance is observed.

Resolving disputes between the stakeholders becomes costly and significantly complex when dealing with monitored performance deficiencies is delayed. Every stakeholder suffers when a project underperforms to the point where disputes cannot be amicably settled. The opposite is an ideal situation; every stakeholder benefits from a project delivered on budget and on time. Determining who is causing the variance and making that party aware of the consequences and remedies is a tough but essential duty of a PM.

Categorical Remedial Action Options

Following the monitoring, evaluation of project status, and determining the root causes of variances, the controlling function of a manager continues with considering what action to take based on the evaluation of planned versus achieved cost data. When the achieved cost data indicates a variance to what has been planned, the PM may choose to:

- **Do not remedy;** look for improvement options
- **Remedy** the root cause and stay with the plan
- **Accept** the variance **and accommodate** it
- **Remedy** the variance **and accommodate** it

Naturally, such decisions are made more so for negative variances of tasks with high cost significance and that pose a potential risk of causing a budget overrun.

In the following paragraphs explaining the details of categorical remedial action options, the definitions of updating and revising budgets are used in the context of their definitions repeated here. *Updating* is incorporating the collected, observed, sorted/processed, and calculated cost data to the WPD and the PCB. If a budget update indicates an overrun (theoretically this case also applies to an underrun), amending the budget to incorporate that overrun is referred to as a budget *revision*. A budget revision requires an approval from a higher authority and a schedule check to verify the impact of that budget revision on project duration. Budget and schedule updates and revisions mandate CF updates and revisions and consequent verification of project finance requirements and costs.

The **"Do not remedy"** option indicates that the subject activity variance is expected to normalize without exceeding its estimated final cost. At early stages of an activity, monitored output can indicate a less than planned value. This observation does not always mean that same output rate will be maintained till that task is completed. The observed output deficiency may be due to the *learning curve effect* which could easily improve as the work advances. Monitoring the periodic output trend will provide the indicators for the PM to decide if any further action will be needed or not.

Searching for possibilities for improving existing performance is one way of defining *continuous improvement*. By monitoring and assessing the progress, the PM, together with the team, is committed to finding areas where the performance can be improved and how it can be improved. This process of continuous improvement takes place regardless of the fact that targeted goals may be accomplished or even exceeded.

As its name implies, this process continues throughout the project independent of any variance to scope, cost, and time variances that may or may not be encountered.

The second corrective action option is to **"remedy"** the root cause of the variance and stay with the plan. In this option, the PM observes and identifies the reasons causing the variance and believes they can be corrected with additional resources that are available. Also, as a result of this variance, the PM concludes that budget and CF revisions are not necessary. This option usually presents itself where there is a negative trend which has not yet turned into a variance that would trigger the threshold requiring a budget revision. The PM addresses the root causes of those variances without having to revise the current plans (i.e., budget and CF).

In the third decision option the PM **"accepts"** the data monitored. This acceptance indicates that the initial estimate is no longer valid. The task quantity and/or unit cost and/or assumed output can no longer be used. Considering the monitored data, the PM will replace them with new values. At the same time, the PM may conclude that a budget revision and/or a *change order* needs to be processed due to the observed and projected variance, the consequences of which need to be schedule and the CF verified.

Budget and consequent schedule and cashflow change could be a revision to be implemented within the organization or it may require an official change order amending the contract between stakeholders.

The act of revising the budget, cashflow, and the schedule is the **"accommodate"** part of this remedial action option. This accommodation may require approvals (and/or blessings if there is no contractual relationship between the parties) from all stakeholders. Obtaining approvals for such accommodations may get complicated as to determining which project stakeholder will be accountable for it. Determining the cause attributable party closely relates to the stakeholder organization and delivery method—contract pricing combination complications as explained later.

Another option for remedial action is to **"remedy"** the cause of the variance by correcting it and **"accommodating"** it within the organization. In this option, the PM has the resources to stop the bleeding without a budget revision or a change order. Most probably due to available contingencies and early warnings of deviation, it is possible to fix the problem and return to target course. The updated budget needs to be schedule and the CF verified.

Remedial Action and Delivery Method/Contract Pricing Complications

The delivery method used for the project, stakeholders' legal status and their contract conditions with each other (i.e., obligations) are the two main factors in making the decision for what corrective action to take. In addition, persuading the party who will pay for the cost variance and/or suffer from a time delay to agree with the proposed remedial actions is a major roadblock in implementing them.

As outlined in Chapter 6 (i.e., Table 6–2), the goals and the priorities of project stakeholders differ substantially depending on the delivery method/contract pricing combination used. The liabilities of the same stakeholders are also governed by the utilized delivery method/contract pricing combination. Accordingly, and subsequently, the party being responsible for increasing cost and/or duration will be accountable for that cost increase and time variance. It is therefore essential to understand which stakeholder will be responsible for any unfavorable variance to cost and duration since that party may resist such changes. It should also be noted that a variance may occur due to a decrease in duration of an activity. A decrease in an activity's duration may not necessarily decrease its cost. Actually, it may increase the activity cost due to overtime and/or additional labor and M/E costs. A shorter activity duration causing a higher activity cost is exemplified by the technique of *crashing critical activity durations* in scheduling by using the Critical Path Method.

Traditionally the designers of projects are not at risk unless a professional misconduct is the issue. Especially in delivery methods where the owner undertakes to carry out the design either in house or outsourced (i.e., any version of DBB), any design-related unfavorable cost and time variance will be the owner's responsibility. Design documents (i.e., drawings, specifications, benchmarks, etc.) that are incomplete, conflicting, contradicting, and indicating conditions that differ at job sites are main causes for delays, lengthy claims, and hefty compensations; all to be endured by the owners and not by the designers whose contracts traditionally protect them from such indemnifications. Another case where the owner is in a position for full delay and/or cost increase compensation is Cost Plus Fee contracts. In this case the owner expects such variances from the start of a project and is prepared for compensation.

Figure 9–5 tabulates stakeholder liabilities for common delivery methods and contract pricing combinations used for regulating the relationships between parties to a project.

Delivery	Contract		Variance Due to	Liability of
	Pricing	Between		
DBB	Unit Price	O and G/C	Errors and Changes	Owner
			Performance	G/C
	Lump Sum	O and G/C	Errors and Changes	Owner
			Performance	G/C
	Lump Sum	O and A/E	Errors, Changes and Performance	Owner
	Lump Sum	O and CM	Errors, Changes and Performance	Owner
	Lump Sum	O and CM@R	Changes	Owner
			Errors and Performance	CM
DB	Lump Sum	O and G/C or JV	Changes	Owner
			Errors and Performance	G/C or JV
	Cost+Fee	O and G/C or JV	Errors, Changes and Performance	Owner
CM	Lump Sum	O and CM@R	Errors, Changes and Performance	CM
	Lump Sum	CM@R and A/E	Errors, Changes and Performance	CM
DBOT	Time till transfer	O and Developer	Errors, Changes and Performance	Developer

Legend: O: Owner G/C: Contractor A/E: Designer
CM: Construction Manager CM@R: CM at Risk
JV: Joint Venture between G/C and A/E
Developer: Entity that designs, builds, and operates the project with its own financing.

Figure 9–5 Stakeholders' major variance liabilities for delivery method and contract pricing combinations.

The design risk is mitigated to the G/C in DB delivery systems where the G/C undertakes design responsibility. While a D/B delivery system may not provide the lowest project cost for the owner, it certainly secures better design coordination and mitigates any design-related delay and cost ramifications.

With the traditional exclusion of designers being at risk for delays and cost increases, owners have generated several contracts that unfairly put G/Cs at risk. This effort results in either unqualified and high risk-taking G/Cs' participation in the bid and/or, qualified G/Cs either declining to bid or bidding inflated prices to compensate for that contract stipulated–perceived risk. A study conducted by the author and Cindy Menches concluded that public impression of owner organization and contract framing techniques can effectively manipulate the contractors' risk perception and consequent bid decisions (Mehmet Nihat Hanioğlu & Cindy Menches (2017): Influence of owner impression and contract framing: could it be used to manipulate construction contractors' perception of risk? *Engineering Project Organization Journal*, DOI:10.1080/21573727.2 017.1305952).

What Is the Right Remedial Action?

The remedial action decision primarily depends on the root cause of variance, the party who caused it, and the contract conditions between the parties. Since the combination of number of stakeholders, number of root causes—even without considering the combinations of root causes—and the contractual obligations between stakeholders easily adds up to a large number, it is very difficult to offer a specific solution for all possible cases. The reader is also cautioned that projects being unique by definition also implies that what might have worked in remedying a specific variance in one project may not be a suitable solution for another.

Figure 9–6 summarizes some generalized remedial actions for typical variance causes.

To find the correct remedy, the PM needs to make the correct diagnosis of the situation. In some cases, allocating additional resources by using contingencies may suffice to complete the project with all the scope, cost, and time requirements satisfied. When remedial action is expected to deplete the foreseen contingencies and require even more resources, there needs to be a more scrutinized analysis of the root cause for this cost overrun. Significant variations that cannot be resolved by using existing contingencies will most likely require approvals from higher management of the organization and/or other stakeholders before additional resources can be allocated and project duration can be extended.

Implementing Remedial Actions

Implementing the remedial action involves almost the same cost planning (i.e., estimating and budget preparation) process. It requires revising the project cost

Variation Cause	Generalized Remedial Action for Variance	
	If Caused by Contractor	If Caused by Owner
Estimating Errors; Scope Creep; Excessive Waste; Damages[1]	Correct error by addressing specific root cause; update or revise budget, schedule, and cashflow.	Issue change order; compensate for additional resources and time; update or revise budget, schedule, and cashflow.
Changes	Issue notice; revise plans.	
Unforeseen Conditions	Revise plans accordingly.	
Performance Issues[2]	Address specific root cause; update or revise budget, schedule, and cashflow.	

Figure 9–6 Generalized stakeholder remedial actions for common variance causes.

Notes:
[1] Damages may be reimbursed by insurance.
[2] Owner's performance issues include delays in approvals and payments.

budget, revising the schedule, verifying both cost budget and duration are acceptable by all stakeholders, and revising the CF and resource schedules accordingly. The revision process as a result of a remedial action and original cost and time planning has two differences:

A. It may require a stakeholder to issue a notice or a change order (i.e., G/C issues the notice and the owner issues a change order). Contract conditions stipulate what needs to happen if a change order request is disputed.
B. Revised plans, both budget and schedule, need to be acceptable to all stakeholders from the resulting scope changes and extended and/or expedited completion duration perspectives.

Once the revisions are approved, they need to be communicated through the entire project organization, which may require briefing and training of staff.

Keeping Records

Keeping records of all updates and revisions together with their supporting documents is a good and multipurpose practice. These documents will not only serve for potential claims and delay analysis, but they will also provide in-depth details for future projects. In addition, it will be a good reference for performance evaluation of PMs and their teams.

Disputes Related to Variances

If the parties to a project cannot agree on who is responsible for a change and which party is to compensate for the cost and time ramifications of that change, that dispute takes the form of a *claim.* The dictionary definition of a claim is *a demand for something due* or, *believed to be due.* In construction project contract administration, a claim is a written form of a demand for cost and/or time compensation. Contracts include articles on how disputes will be settled. Amicable settlement is the fastest and least expensive option. A more time-consuming and expensive alternative is arbitration, and litigation is the most expensive and longest option.

Some contracts mandate work to be continued even though the claimed cost compensation is significant. Entering such a contract is extremely risky since the contractor has to finance not only the arbitration/litigation of the claim but also the work itself until the claim is settled, which may come months if not years after project completion.

If the contract allows, one option the contractor may consider is to stop work until the claim is settled. Stopping work is not an easy decision since it will not only have a big impact on the project cost and duration but it will also deteriorate the relationship between project stakeholders.

Disputes and especially claims may produce devastating results if they are not properly handled. The best practice for dealing with disputes is avoiding them. This can be accomplished by distributing the project risk fairly among the stakeholders in contracts and by approaching disputes with an open mind. Eliminating all variances is not practically possible. Risks generate variances and variances generate more risks in return. Equitable sharing of unavoidable project risks and resolving disputes fairly and without delay is the key to completing projects within budgets and on time.

Cost Control Process Schematics

Cost control provides current and projected indicators that are essential for any PM to keep their projects on cost and time. It incorporates accomplished quantities, unit costs and efficiencies to be compared to those planned and produces feedback. It collects data from accounting and scheduling and combines that data with other factors such as scope, changes and unforeseen condition to project cost and physical work projections.

Figure 9–7 presents the schematics of the cost control process.

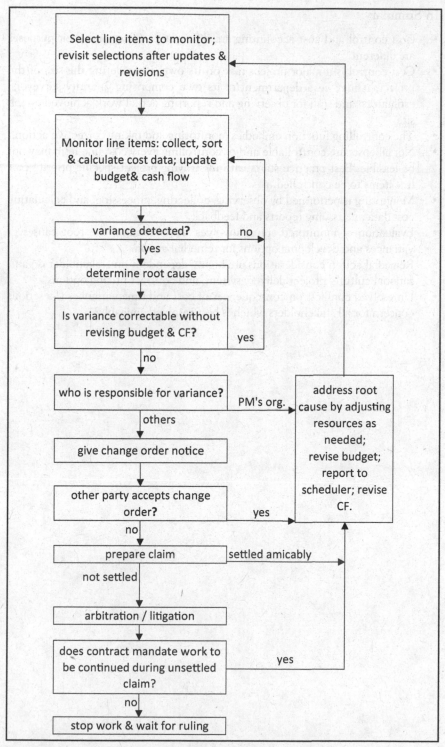

Figure 9–7 Cost control flowchart for contracted (outsourced) projects.

In Summary

- Cost control and cost accounting are different practices and their purposes are different.
- Cost control is a major project task on its own. Considering the size of the project, it may be a department of its own comprising quantity surveyors, estimators, and staff for observing and reporting actual work achieved on job sites.
- The controlling function embodies monitoring and taking corrective action.
- Not all costs are controllable and/or controlling every cost line item may not be feasible. Best practice starts with identifying project-specific priority cost line items to be controlled.
- Monitoring is performed by observing, collecting, processing, and calculating cost data and issuing reports and feedback.
- Evaluation of monitored status involves understanding the root causes of variances and developing options for remedial action.
- Remedial action considerations are highly dependent on stakeholder organizational culture, project delivery system, and contract pricing options.
- Unresolved conflicts on consequences of cost and time variances is a serious concern for all stakeholders which is best resolved amicably.

Appendixes

Appendix A: Cost Budget Forms for Hotel Project Stakeholders

Summary

Description	Initial Budget	Apprvd. Changes	Adjusted Budget	Committed Costs	Pending Costs	Cost @ Completion	Over/ Under
1 DEVELOPMENT COST							
1.1 Project Management Costs							
1.2. Owning Company Costs							
1.3. Licenses, Permits, and Fees							
1.4 Finance Costs							
Total Development Cost							
2 LAND							
2.1 Purchase/Lease							
2.2 Infrastructure and Improvements							
Total Land Cost							
3 CONSTRUCTION AND FIT-OUT							
3.1 Design Costs							
3.2 Construction Costs							
3.3 FF&E							
Total Const. and Fit-Out Cost							
4 PRE-OPENING AND WORKING CAPITAL							
4.1 Pre-opening							
4.2 Working Capital							
Total Pre-Opening & Working Capital Cost							
5 CONTINGENCIES							
5.1 Development							
5.2 Construction and Fit-Out							
5.3 Pre-opening							
5.4 Escalation							
Total Contingencies							
TOTAL PROJECT COST							

APPENDIX A.1: FOR OWNERS

1. Development Costs

Description	Initial Budget	Approved Changes	Adjusted Budget
1.1 PROJECT MANAGEMENT COSTS			
1.1.1 Project Manager			
1.1.1.1 salary and benefits			
1.1.1.2 housing, car, and schooling			
1.1.1.3 relocation and home leave			
1.1.2 Assistant Project Manager (salary only)			
1.1.3 Engineer (salary only)			
1.1.4 FF&E Manager			
1.1.4.1 salary			
1.1.4.2 benefits, housing, car			
1.1.4.3 relocation and home leave			
1.1.5 Support Staff (salary only)			
1.1.6 Travel and expenses			
1.1.7 Contingency			
Total Project Management Cost	_____	_____	_____
1.2 OWNING COMPANY COSTS			
1.2.1 Managing Director			
1.2.2 Finance Director			
1.2.3 Support Staff (accountants, secretaries)			
1.2.3.1 Accounting			
1.2.3.2 Secretaries			
1.2.3.3 Staff for direct supplier payments			
1.2.4 Office costs			
1.2.4.1 rent and utilities			
1.2.4.2 telephone and courier			
1.2.4.3 supplies			
1.2.5 Direct Site Expenses			
1.2.5.1 Site hoarding			
1.2.5.2 Site survey, soil tests			
1.2.5.3 Utility connection and works			
1.2.6 Public Utility fees and expenses			
1.2.6.1 Sewerage			
1.2.6.2 Electrical			
1.2.6.3 Boulevard			
1.2.6.4 Water, telephone, miscellaneous			
1.2.7 Pre-development expenses			

(Continued)

Appendix A.1 (Continued)

Description	Initial Budget	Approved Changes	Adjusted Budget
1.2.8 Consultant fees (non-construction)			
1.2.8.1 Legal counsel			
1.2.8.2 Accounting			
1.2.8.3 Public Relations, insurance, etc.			
1.2.9 Travel and related expenses			
Total Owning Company Costs	___	___	___
1.3 LICENSES, PERMITS, AND FEES			
1.3.1 Architect of Record			
1.3.2 Technical Service fees			
1.3.3 Purchasing Agent fee			
1.3.4 Customs Clearance brokerage fees			
1.3.5 Zoning, planning permits			
1.3.6 Operation/Occupancy permits			
Total Licenses and Permits	___	___	___
1.4 FINANCE COSTS			
1.4.1 Loan commission fees			
1.4.2 Appraisal fees			
1.4.3 Lenders' legal fees			
1.4.4 Commitment fees			
1.4.5 Stamp duty, transfer charges			
1.4.6 Interest during construction			
1.4.7 Financial institution supervision requirements			
Total Finance Costs	___	___	___
TOTAL DEVELOPMENT COST	___	___	___

2 Design Costs		
	FEES	*REIMBURSABLES*
2.1 MASTER PLANNING		
If not included in Owning		
Company Costs		
subtotal	_____	_____
2.2 CONCEPT ARCHITECT		
Base fees		
Red-line review		
Marketing documents		
Reimbursables		
subtotal	_____	_____
2.3 LANDSCAPE ARCHITECT		
Base fees		
Reimbursables		
subtotal	_____	_____
2.4 INTERIOR DESIGN		
Base fees		
Lighting fee		
Renderings		
Reimbursables		
subtotal	_____	_____
2.5 MEP DESIGN		
Base fees		
Redline review		
Reimbursables		
subtotal	_____	_____
2.6 STRUCTURAL ENGINEER		
Base fee		
Reimbursables		
subtotal	_____	_____
2.7 KITCHEN AND LAUNDRY		
Base fee		
Reimbursables		
subtotal	_____	_____
2.8 SIGNAGE, GRAPHICS, AND		
UNIFORM		
Base fee		
Reimbursables		
subtotal	_____	_____
2.9 DESIGN CONTINGENCY		
TOTAL DESIGN COST	_____	

Appendix A.1 (Continued)

	Description	Initial Budget	Approved Changes	Adjusted Budget
3 Construction and Fit-Out Costs				

3.1 Design costs: see separately provided section.

3.2 CONSTRUCTION COSTS
   3.2.1 **CIVIL AND STRUCTURE**
      3.2.1.1 Site Prep. and Infrastructure
      3.2.1.2 Site Hardscape
      3.2.1.3 Landscape and Irrigation
      3.2.1.4 Building Structure
   3.2.2 **ARCHITECTURAL**
      3.2.2.1 Building Envelope
      3.2.2.2 Interior Finishes
      3.2.2.3 Millwork
   3.2.3 **MEP**
      3.2.3.1 HVAC
      3.2.3.2 High Voltage System
      3.2.3.3 Low Voltage System
      3.2.3.4 Plumbing
      3.2.3.5 Fire Protection
      3.2.3.6 Lifts and Escalators
   3.2.4 **Miscellaneous Indirect Costs**
      3.2.4.1 Permits and Fees
      3.2.4.2 Surveys and Control
      3.2.4.3 Tests and Inspections
      3.2.4.4 Mock-Ups
      3.2.4.5 Owner's PM Office
      3.2.4.6 Training

3.3 F&E
   3.3.1 Furniture and Furnishings
   3.3.2 Major Equipment
   3.3.3 Other Equipment

Taxes: Included in costs

APPENDIX A.2: FOR CONTRACTORS

Construction, Design, and FF&E Direct Costs				
Description	*Unit*	*Qty.*	*Unit Cost ($)*	*Cost ($)*
1 SITE PREPARATION AND INFRASTRUCTURE				___
1.1 Site Preparation				
1.2 Infrastructure Systems				
1.3 Site Access				
1.3.1 Vehicle Entry				
1.3.2 Parking areas				
1.3.3 Other site access works				
2 SITE DECORATION				___
2.1 Decorative Hardscape				
2.2 Pools/Slides				
2.3 Water Features/Fountains				
2.4 Site Furnishings				
3 LANDSCAPE AND IRRIGATION				___
3.1 Irrigation System				
3.2 Fine Grading and Soil Preparation				
3.3 Plants and Other Landscape Elements				
3.4 Maintaining Landscaped Areas				
4 BUILDING STRUCTURE				___
4.1 Foundation				
4.2 Superstructure				
4.2.1 Form Work				
4.2.1.1 Walls				
4.2.1.2 Columns				
4.2.1.3 Slabs				
4.2.1.4 Special Forms				
4.2.2 Reinforcing Steel				
4.2.3 Cast-In-Place Concrete				
4.2.4 Precast Elements				
4.2.5 Structural Steel				
4.2.6 Structural Timber				
5 BUILDING ENVELOPE/PERIMETER				___
5.1 Roof System				
5.2 Prefabricated Exterior Walls				
5.3 Blockwork				
5.4 External Wall Insulation				
5.5 Façade Systems				
5.6 Plaster Rendering				
5.7 Windows				
5.8 Skylights				
5.9 Exterior Paint				
5.10 Louvres and Screens				
5.11 External Doors				
5.12 Other Perimeter/Envelope Finishes				

(Continued)

Appendix A.2 (Continued)

Description	Unit	Qty.	Unit Cost ($)	Cost ($)
5.13 Exterior Building Signage				
5.14 Exterior Building Lighting				
6 INTERIOR FINISHES				—
6.1 Internal Partitions				
6.2 Ceilings				
6.3 Plaster Rendering				
6.4 Floor Finishes				
6.5 Wall Finishes				
6.6 Ceiling Finishes				
6.7 Internal Doors (including frames)				
6.7.1 Wood Doors				
6.7.2 Metal Doors				
6.7.3 Internal Glass Doors				
6.7.4 Specialty Doors				
6.8 Ironmongery				
6.8.1 Internal Doors				
6.8.1.1 Wood Doors				
6.8.1.2 Metal Doors				
6.8.1.3 Glass Doors				
6.8.1.4 Specialty Doors				
6.8.2 External Doors				
6.8.3 Ironmongery for Windows				
6.8.4 Other ironmongery				
6.9 Bathroom/WC Accessories				
6.9.1 Guest Rooms				
6.9.2 Suites				
6.9.3 Public Areas				
6.9.4 Back of House				
6.10 Specialty Systems				
6.11 Interior Signage				
6.12 Interior Landscape				
6.13 Lift Interior Decoration Allowance				
7 MILLWORK				—
7.1 Floors, Platforms, and Tiers				
7.2 Wall, Columns, and Partitions				
7.3 Ceilings				
7.4 Counters				
7.5 Cabinets				
7.5.1 Guest rooms/Suites				
7.5.2 Public Areas				
7.5.3 Back of House Cabinets				
8 HEATING, VENTILATION AND AIR-CONDITIONING				—
8.1 Fan Coil Units				
8.2 Ventilation Fans				
8.3 Exhaust Fans				
8.4 Kitchen Exhaust Fans				
8.5 Fire Dampers				
8.6 Volume Air Dampers				

(Continued)

Description	Unit	Qty.	Unit Cost ($)	Cost ($)
8.7 Air Handling Units				
8.8 Water Cooled Units				
8.9 Refrigeration Plant				
8.10 Chemical Treatment System {refrigeration plant}				
8.11 Boiler Plant				
8.12 Ductwork and Distribution				
8.13 Piping, Valves, and Fittings				
8.14 Controls				
8.15 Test and Commissioning (HVAC Systems)				
9 HIGH VOLTAGE SYSTEM				—
9.1 Generator-Set				
9.2 Transformers				
9.3 High Voltage Distribution Boards				
9.4 High Voltage Metering				
9.5 Sub Distribution Boards				
9.6 Installation and Cabling				
9.7 Light Fixtures				
9.8 Dimming System/Controls				
9.9 Emergency Lights and Signage				
10 LOW VOLTAGE SYSTEM				—
10.1 Telecom System				
10.2 Audio Visual Systems				
10.3 MATV System				
10.4 Interactive TV System				
10.5 Security System				
10.6 UPS				
10.7 Computer System				
10.8 Electronic Door Locks				
10.9 BAS				
11 PLUMBING				—
11.1 Cold Water System				
11.2 Hot Water System				
11.3 Sewage Wastewater System				
11.4 Swimming Pool/Water Features				
11.5 Gas System				
11.6 Drainage				
11.7 Kitchen and Valet				
11.8 Sanitary Ware and Fixtures				
11.8.1 Guest rooms, Sanitary Ware				
11.8.2 Guest room, Sanitary Faucets and Fittings				
11.8.3 Public Areas, Sanitary Ware				
11.8.4 Public Areas Sanitary Faucets and Fittings				
11.8.5 Other Sanitary Ware				
11.8.6 Other Sanitary Faucets and Fittings				
11.9 Utility Metering				

(Continued)

Appendix A.2 (Continued)

Description	Unit	Qty.	Unit Cost ($)	Cost ($)
12 FIRE PROTECTION				___
12.1 Fire Protection Equipment				
12.2 Fire Alarm Detection				
12.3 Test and Commission				
13 LIFTS AND ESCALATORS				___
13.1 Passenger Lifts				
13.2 Service Lifts				
14 INDIRECT SITE COSTS				___
14.1 Permits and Fees				
14.2 Survey and Control				
14.3 Test and Inspections				
14.4 Mock-up Room Construction and Assembly				
14.5 Site Facilities for Owner's Project Manager				
14.6 Preliminaries				
14.7 Cost of 90 days pre-contract period				
15 FF&E FEES				___
15.1 Purchasing Agency				
15.2 Handling and Storage				
15.3 Installation				
16 DESIGN FEES				___
16.1 Architectural				
16.2 Structural				
16.3 MEP				
16.4 Interior				
16.5 Lighting				
16.6 Landscape				
16.7 Kitchen and Valet				
16.8 Others (specify)				
TOTAL				

APPENDIX A.3: FOR HOTEL OPERATORS

Hotel Operator's Project Cost Budget			
Description	*Initial Budget*	*Approved Changes*	*Adjusted Budget*
1 PRE-OPENING EXPENSES			
1.1 Payroll and Related			
1.1.1 Rooms			
1.1.2 Food and Beverage			
1.1.3 Other Departments			
1.1.4 Admin and General			
1.1.5 Human Resources			
1.1.6 Sales and Marketing			
1.1.7 POMEC			
1.2 Pre-opening Office Expenses			
1.2.1 Rent			
1.2.2 Utilities and Maintenance			
1.3 Support Teams			
1.4 Training Expenses			
1.5 Opening Ceremony			
1.6 Contingency			
Total Pre-opening Expenses	____	____	____
2 WORKING CAPITAL			
2.1 Initial Inventory			
2.2 Consumables			
2.3 Cash Float			
Total Working Capital	____	____	____
3 FURNITURE, FURNISHINGS, AND EQUIPMENT			
3.1 **Furniture and Furnishings**			
3.1.1 Guest Rooms			
3.1.2 Public Areas			
3.1.3 Art Program			
3.1.4 Interior Signage			
Total Furniture and Furnishings	____	____	____
3.2 **Major Equipment**			
3.2.1 Kitchens and Bars			
3.2.2 Laundry			
Total Major Equipment			
3.3 **Operating Equipment**			
3.3.1 Silverware			
3.3.2 Chinaware			
3.3.3 Glassware			
3.3.4 Linen			
3.3.5 Uniforms			
Total Operating Equipment	____	____	____

(Continued)

Appendix A.3 (Continued)

Description	Initial Budget	Approved Changes	Adjusted Budget
3.4 Special Equipment			
3.4.1 Management System			
3.4.2 Office Equipment			
3.4.3 Material Handling Trucks			
3.4.4 Cleaning Equipment			
3.4.5 Dining Room Wagons			
3.4.6 Shelving and Lockers			
3.4.7 Vehicles			
3.4.8 Banquet Equipment			
3.4.9 Recreational Equipment			
3.4.10 Guest Room Accessories			
Total Special Equipment	_____	_____	_____
3.5 Miscellaneous Equipment, Tools, and Utensils			
3.5.1 Kitchen and Stewart Utensils			
3.5.2 Dining Room Accessories			
3.5.3 Engineering Tools and Equipment			
3.5.4 Housekeeping Utensils			
3.5.5 Miscellaneous Equipment			
Total Miscellaneous Equipment, Tools, and Utensils	_____	_____	_____
Total FF&E	_____	_____	_____

APPENDIX A.4: DESIGN WBS

Property Development Project Typical Design WBS (Phases & Details)

1. **Concept Design**
   1.1 Design Statements
   1.2 Concept/Layout
   1.3 Architectural
   1.4 Interior
   1.5 Site Decorations and Landscape

2. **Schematic Design**
   2.1 Architecture Schematics (1:200)
   2.2 Phase 1 Interiors (1:100)
   2.3 Structure
   2.4 MEP Systems
   2.5 Vertical Transportation Study
   2.6 Site Decorations and Landscape

3. **Design Development**
   3.1 Architecture
      3.1.1 Building Interior:
         3.1.1.1 1:100 Drawings
         3.1.1.2 1:50 Details
      3.1.2 Building Façade and Exteriors:
         3.1.2.1 1:100 Drawings
         3.2.1.2 1:50 Details
      3.1.3 Specifications
   3.2 Interior Design
      3.2.1 Phase II Interiors (1:50)
      3.2.2 F.F. and E. Interiors Plan
      3.2.3 Interior Details:
         3.2.3.1 Color Renderings
         3.2.3.2 Outline Specifications
         3.2.3.3 Millwork Profiles
         3.2.3.4 Ironmongery/Door Schedule
         3.2.3.5 Toilet Fixture/Accessories
      3.2.4 Mockup Room:
         3.2.4.1 Plans
         3.2.4.2 Details
         3.2.4.3 FFE
         3.2.4.4 Millwork
      3.2.5 Specifications
   3.3 Structure
      3.3.1 1:100 Drawings
      3.3.2 1:50 Details
      3.3.3 Specifications
   3.4 MEP (Incl. HVAC, LV, HV, and Plumbing)
      3.4.1 Equipment Schedules
      3.4.2 Equipment Locations
      3.4.3 Specifications
      3.4.4 Recommended Manufactures
      3.4.5 Single Line Mechanical (1:100)
      3.4.6 Mechanical Riser Diagrams
      3.4.7 Vertical Riser Layout

(Continued)

Appendix A.4 (Continued)

Property Development Project Typical Design WBS (Phases & Details)

   3.4.8 Ducting Runs
   3.4.9 Cooling Tower Piping
   3.4.10 Guestroom Vertical Riser (1:50)
   3.4.11 Typical Guestroom Electrical (1:50)
   3.4.12 Main Distribution Board Diagrams
   3.4.13 Vertical and Horizontal Runs
   3.4.14 Building Automation System
   3.4.15 Fire Protection Systems
   3.4.16 Fire Alarm System
   3.4.17 Low Voltage Systems
   3.4.18 Computer Systems Schedule
   3.4.19 Lighting Layout and Fixture Schedule
   3.4.20 Water Main, Distribution, and Storage
   3.4.21 Water Softening and Treatment
   3.4.22 Kitchen and Laundry Equipment List, Layout and Specs.
  3.5 Vertical Transportation Drawings and Specs.
   3.5.1 Guest Elevators
   3.5.2 Service Elevators
   3.5.3 Escalators
  3.6 Site Decoration and Landscape
   3.6.1 Sections/Elevations
   3.6.2 Planting Schedule
   3.6.3 Specifications
4. FF&E
 4.1 Furniture and Furnishings
 4.2 Major Equipment
 4.3 Operating Equipment
 4.4 Specialty Hotel Equipment
 4.5 Miscellaneous Equipment, Tools, and Utensils
 4.6 Specifications and Recommended Suppliers
5. Bid/Construction Documents

Appendix B: Sample Work Package Dictionary Form

This sample WPD form is filled with the information obtained at Step 2 of Hotel Building Superstructure estimating example of Chapter 6.

HOTEL DEVELOPMENT PROJECT TASK DICTIONARY *(page 1 / 2)*

Task: 3.2.4.2.1. Building Superstructure, Slab for Floor 1

Scope:
Reinforced concrete for Floor Slab 1 of the 10 floor hotel building superstructure. Task includes formwork, scaffolding, reinforcing steel, electrical, mechanical and plumbing insets/sleeves, anchors for façade elements and placing concrete and finishing it as designed and specified contract documents.

Duration:
10 days from erecting formwork and scaffolding to stripping and preparing them for the next use.

Schedule:
Starts 1 day after the the concrete placement to the prior structural element, 3.2.4.1. Basement Slab.

Milestones:
No specific milestone but the target completion is 120 days for all superstructure tasks.

Assumptions:
a. All floor slabs are structurally identical and one reinforced concrete takeoff will sufficiently represent the work for all of them.
b. Ready mix concrete from a local batch plant will be used. The price for concrete includes transportation, pump and tests.
c. Adjustable metal props with beam drop heads to accommodate 4"x4" wood beams will be used to support the ¾" plywood deck.
Column formwork will be prefabricated plywood/2x4 panels. The same panels will be used for shear wall around elevator shaft. 2x8" wood planks will be used to edge slab perimeter and voids.
d. The deck will be extended 6' in all directions to provide working and safety area.
e. All reinforcement will be delivered cut and bent to size.
f. Site installations, staging and storing areas, tower crane, on site transportation and other services will be used.
The cost of these services will not be included in the cost of the task; site services cost will be distributed over project tasks as a general overhead mark-up.

Method:
a. Formwork starts simultaneously with columns, shear wall and props-beams and followed with deck.
b. Mechanical, electrical and plumbing (MEP) insets are placed.
c. Reinforcement and anchors tied in place.
d. For the pouring day, site access to truck mixers and concrete pump will be cleared to start pouring from the SE corner.
e. Poured concrete will be compacted by electrical and diesel vibrators. After manual levelling, concrete will be finished by power trowels.

(Continued)

Appendix B (Continued)

HOTEL DEVELOPMENT PROJECT TASK DICTIONARY

Task: 3.2.4.2.2. Building Superstructure, Slab for Floor 1 (page 2/2)

Progress Measuring Metrics:
Approximated cost based weights (i.e. percentages) will be used to measure completion progress:
Building excluding FF&E is [. . .] (*) % of Building with FF&E (excluding land);
Structure is [. . .] (*) % of Building (excluding FF&E)
Superstructure is [. . .] (*) % of Structure;] (*)
Each slab is 10% of Superstructure;
Forms of one slab is 40% of that slab;
Rebar of one slab is 25% of that slab;
Concrete of one slab is 35% of that slab.
(*) These percentages will be filled out when estimates for these tasks are completed.

Cost: $ 206,480
See attached takeoff and BoQ sheets for details (Tables 6.2 & 6.5).
The general site overhead and head office overhead is excluded in the cost.

Project Manager in Charge: Ms. Solorio

Attachments: Takeoffs and BOQ.

Appendix C: Typical Progress Report Forms

APPENDIX C.1: SAMPLE DAILY REPORT

DAILY REPORT			
Project:		date:	
		report no:	

PRODUCTION

		quantities	
WBS code	description	target	actual

MATERIALS

description	stock	issues/remarks

LABOR

Resource		hours		worked on WBS code
trade code	name	regular	overtime	

MACHINERY /EQUIPMENT

M/E		hours		worked on WBS code
M/E code	description	regular	overtime	

ISSUES/REMARKS

prepared by:

Sample daily report format for a construction site. Depending on the size and number of activities worked on, the reporting manager may have to add pages or separately report production, materials, labor, and M/E sections.

APPENDIX C.2: MONTHLY REPORTS

Project Manager's Monthly Report

NO [..]

CONTENTS

1. EXECUTIVE SUMMARY
2. PROJECT FACT SHEET
3. CONSTRUCTION COSTS—BUDGET
4. CONSTRUCTION PROGRESS PAYMENTS
5. CONSTRUCTION COSTS—CASH FLOW
6. LABOR—HOURS ON CONSTRUCTION
7. PROJECT WORK SCHEDULE
8. CONCLUSION/REMARKS
9. PHOTOS

Date

(Continued)

1. **EXECUTIVE SUMMARY** (includes the summary of sections 3, 4, 5, 7, and 8)
2. **PROJECT FACT SHEET**

PROJECT DATA

Number of Keys   :
Number of Bays   :
Total Area   :
Hotel area   :
Office Area   :
Parking Area   :

DESIGNERS

Design Architect   :
Local Architect   :
Interior Designer   :
Structure   :
MEP   :
Lighting   :
Landscape   :

CONSTRUCTION DATA

General Contractor   :
Contract date   :
Contract end date   :
Time extension   :
Adjusted end date   :
Scheduled % completion :
Actual % completion   :

PROGRESS PAYMENTS

	Contract	*Variations*	*Adjusted*	*Achieved*	*% Achieved*
Design Architect					
Local Architect					
Interior Designer					
Structure					
MEP					
Lighting					
Landscape					
Gen. Contractor					
FF&E					

(*Continued*)

Appendix C.2 (Continued)

3. Construction Costs—Budget

Description	Initial Budget	Apprved. Changes	Adjusted Budget	Commitd. Costs	Pending Costs	Cost @ Compl.	Over / (Under)
1. Contracted Cost							
1.1 Civil and Structure							
1.2 Architectural							
1.3 MEP							
1.4 Contractor Indirect Costs							
1.5 Design Fees							
1.6 Escalation Allowance							
Subtotal Contracted Cost							
2. Hotel FF&E							
2.1 Furniture and Furnishings							
2.2 Major Equipment (K&L)							
2.3 Equipment							
2.4 Contingency							
Subtotal Hotel FF&E							
TOTAL CONSTRUCTION COST							

4. Construction Progress Payments

CONTRACT SUM :

ADVANCE PAID :

APPROVED VARIATIONS :

ADJUSTED CONTRACT SUM :

CLAIMED TO DATE :

LESS RETENTION :

ADVANCE RETURN DEDUCTION :

OVERALL DUE :

LESS PREVIOUS PAYMENTS :

PAYMENT DUE BEFORE TAXES :

TAXES :

PAYMENT DUE TOTAL (INCL. TAXES) :

PAYMENT DUE DATE :

5. CONSTRUCTION COSTS—CASH FLOW *[insert graphs]*

(Continued)

6. **DIRECT LABOR and M/E/ HOURS** *[insert graphs]*

7. **SCHEDULE** *[insert graph]*

8. **CONCLUSION—REMARKS**

   Design
   Construction
   FF&E
   Licenses/Permits
   Overall Progress
   Payments
   Outstanding Issues
   Other Remarks

9. **PHOTOS**

About the Author

Mehmet Nihat Hanioğlu graduated from Middle East Technical University, Ankara, Turkey with a BS in Civil Engineering. He took part in heavy construction projects in Iraq and Saudi Arabia, working his way up in construction project management. After returning to his home base in Istanbul, he started an affordable housing development company that made its way to the top 10 in the nation. He also owned and managed an international construction claims and project management consulting company.

After moving to Chicago, he obtained his MS and PhD in Construction Engineering and Management at Illinois Institute of Technology, Armour College of Engineering. Along with his academic studies, he represented international design and development projects for the hospitality brands Hyatt and Wynn Resorts. He has overseen new design/build, repurposing, and major renovation projects in Europe, North and South America, Asia, and the Middle East.

In 2010, he began sharing his lifelong learnings with emerging engineers and project managers as an instructor in the City University of New York Lehman College and later held teaching positions at IIT and California State University at Sacramento.

He currently lives in Arizona with his wife Kathleen and their cavapoo girl Ginger and is working on his golf game and drumming.

Index

Printed in the United States
by Baker & Taylor Publisher Services

Printed in the United States
by Baker & Taylor Publisher Services